SEVENTH CATALOG OF THE
Vascular Plants of Ohio

This work was sponsored by the

Ohio Department of Natural Resources.

Publication costs were partially underwritten by

the ODNR's Division of Natural Areas and Preserves.

Tom S. Cooperrider, Allison W. Cusick,
and John T. Kartesz, editors

SEVENTH CATALOG OF THE
Vascular Plants of Ohio

OHIO STATE UNIVERSITY PRESS Columbus

Cover illustration. *Trillium grandiflorum*, Large White Trillium, State Wildflower of Ohio. Sketch by John Cooperrider.

Source for map on p. x: Miami University, Oxford, Ohio.

Library of Congress Cataloging-in-Publication Data

Seventh catalog of the vascular plants of Ohio / Tom S. Cooperrider, Allison W. Cusick, and John T. Kartesz, editors.
    p.   cm.
  Includes bibliographical references (p.) and indexes (p.)
  ISBN 0-8142-0858-4 (alk. paper)—ISBN 0-8142-5061-0 (pbk. : alk. paper)
    1. Botany—Ohio—Classification. 2. Botany—Catalogues and collections—Ohio   I. Cooperrider, Tom S.   II. Cusick, Allison W.   III. Kartesz, John T.
  QK180 .S48 2001
  581.9771′01′2—dc21

                                          2001021041

Text and jacket design by Diane Gleba Hall.
Type set in Goudy by G&S Typesetters.
Printed by Thomson-Shore, Inc.

9 8 7 6 5 4 3 2 1

This book is dedicated to all the collectors,

whether of many specimens or few,

who over the past two hundred years

have documented the plant life of Ohio,

> *and*

to all the herbarium curators, assistants,

collection managers, technicians, and volunteers

who have maintained and cared for the specimens,

preserving a part of Ohio's botanical heritage.

FOREWORD

Plants are an integral part of our everyday lives. We live in a world that surrounds us with plants, from lawns and roadsides to agricultural fields and natural landscapes. Most of us come into contact with plants, including trees, wildflowers, and alien weeds, on a daily basis. Among the numerous benefits plants provide are food, fiber, shelter, medicines, and most important, oxygen. They also clean the air, provide shade, control soil erosion, and add diversity and beauty to our lives. Human nature being what it is, at some point we inevitably become curious about our surroundings and seek to know the names of those plants that have so frequently caught our eye, or those that are new and strange to us.

There are several excellent field guides, manuals, and other reference works available today for identifying plants. However, establishing the identity of a plant can be challenging for professional and amateur alike. The first step in the process of identification is to narrow the field of probabilities to a list based on what has been previously inventoried. This is the starting point, and that is what the *Seventh Catalog of the Vascular Plants of Ohio* enables us to do—narrow our search.

The Ohio Department of Natural Resources (ODNR), Division of Natural Areas and Preserves, does a very good job of inventorying and tracking Ohio's rarest species of vascular plants. The same cannot be said, however, for the more common species that constitute the overwhelming majority of our flora. The ODNR regularly receives requests for a listing of all the vascular plants of the state. Until now, a modern comprehensive catalog of the vascular flora of Ohio for the use of professional and amateur botanists, naturalists, and conservationists has been regrettably absent.

The new *Seventh Catalog,* orchestrated by Tom S. Cooperrider, Professor Emeritus of Biological Sciences at Kent State University, and until recently Chair of the Ohio Flora Committee of The Ohio Academy of Science, provides us with just such a comprehensive account of the vascular flora as it is known today. The taxa are placed in a modern classification system and given contemporary nomenclature. With scientific names, synonyms, common names, and two thorough indexes, this is a user-friendly reference for conservation and environmental workers in Ohio as well as those in neighboring states.

Prepared by Cooperrider and a team of distinguished colleagues, coeditors Allison W. Cusick and John T. Kartesz, and coauthors Barbara K. Andreas, Cusick, John V. Freudenstein, and John J. Furlow, this book revives an important series, one that began with John Strong Newberry's first *Catalogue* in 1860. It is also a splendid and extremely useful supplement to the four Vascular Flora of Ohio volumes published by the Ohio State University Press. These volumes resulted from the work of the Ohio Flora Committee, which began in 1950 under the leadership of the late E. Lucy Braun.

GUY L. DENNY, Chief (retired)
ODNR Division of Natural Areas and Preserves

CONTENTS

Map of Ohio and counties

INTRODUCTION

Tom S. Cooperrider

The plants growing wild in Ohio—from the trees in an old-growth forest to the weeds in a parking lot—affect many facets of the lives of Ohio's citizens, among them our science, aesthetics, recreation, environment, health, and economy. As we complete some two hundred years of work on the Ohio flora and as the bicentennial, in 2003, of Ohio's statehood approaches, it seems an appropriate time for an accounting of the state's flora as it is known today. The catalog lists some 3,000 taxa of vascular plants that have been collected in Ohio.

Overview of the Flora

In the late 1700s, an old-growth forest composed of approximately one hundred species of broad-leaved deciduous trees and a few species of needle-bearing evergreens covered nearly all the Ohio region. In later life, recalling his childhood in the 1840s, Morris Schaff (1905) left a firsthand description of the original forest in Licking County, Ohio. He, like other observers, remembered how far it was from the ground to the first limbs.

> When I was a boy three fourths of Etna Township was covered by a noble primeval forest. And now, as I recall the stately grandeur of the red and white oaks, many of them six feet and more in diameter, towering up royally fifty and sixty feet without a limb; the shellbark hickories and the glowing maples, both with tops far aloft; the mild and moss-covered ash trees, some of them over four feet through; the elms and sturdy beeches, the great black walnuts and the ghostly-robed sycamores, huge in limb and body, along the creek bottoms, I consider it fortunate that I was reared among them and walked beneath them . . .

The forest was home also to a number of shade-tolerant shrubs, vines, and ferns living beneath the trees. And during the early part of the growing season, in April and May, a host of about 200 species of wildflowers appeared on the forest floor, most completing their flowering cycle before the leaves expanded above them, cutting down the light.

The first section of this Introduction is adapted from Cooperrider (1999).

Ninety-five percent of Ohio was forested (Dean and Chadwick, 1940). The other five percent was open and treeless: prairie patches, glacial lakes, acid bogs, alkaline fens, open fields of sand, open marshes, rocky cliffs and ledges, and rivers and their muddy banks, each habitat with a distinctive flora. In all habitats combined, both wooded and open, the native flora included about 1,800 species of vascular plants.

The wooded land in Ohio was part of a more extensive deciduous forest that covered much of the eastern half of the continent (Braun, 1950). One imagines that the Native Americans, whose ancestors had lived there for centuries, would have thought the forest a constant, an element of nature endless in time. But the settlers of the early nineteenth century set out to clear the land for farming, and they succeeded. The trees were cut and most simply burned at or near the site where they fell (Knepper, 1976). Today a few remnants of the original forest as well as remnant areas of the various nonwooded habitats survive, giving us a glimpse here and there of primeval Ohio.

Scores of alien plants, mostly Eurasian in origin, moved rapidly into the disturbed habitats left by the clearing. Since then, alien plants have continued to enter the flora and, by latest count of herbarium specimens, some 900 species of nonnative vascular plants have been collected in Ohio growing outside cultivation. As noted below, about 500 of these species have become naturalized, that is, firmly established and maintaining themselves in the flora; the other 400 have thus far been ephemeral.

Study of the Flora

Naturalists of the late eighteenth and early nineteenth centuries were the first to make herbarium specimens documenting the Ohio flora (Stuckey, 1984). As botanical work continued during the first half of the nineteenth century, the major components of the state's flora were gradually determined (Lowden, 1997; Stuckey, 1984). In 1860, John S. Newberry published a general catalog of Ohio's vascular plants listing 1,276 species regarded as native and 101 thought to be alien (Cooperrider, 1992). A series of such catalogs followed; the second was written by Henry C. Beardslee (1874), the third by William A. Kellerman and William C. Werner (1893), the fourth by Kellerman (1899), and the fifth and sixth by John H. Schaffner (1914, 1932). The goal of each catalog was to account for the known flora of its day and—beginning with the second—to add new taxa discovered since the publication of the previous catalog and to delete taxa learned to have been attributed to the state's flora in error.

This seventh catalog continues the series, incorporating the additions and changes resulting from the field and herbarium work on Ohio plants done by botanists in the past seven decades. Several significant events have occurred during this period.

In the early 1950s, E. Lucy Braun organized and directed the work of the Ohio

Flora Committee of The Ohio Academy of Science. The endeavor came to be called the Ohio Flora Project (Cooperrider, 1984). The goal of the project was to produce several books that would include identification keys, illustrations, commentary, and county dot distribution maps for each species. The books that have appeared to date include two by Braun (1961, 1967), one by T. Richard Fisher (1988), and one by myself (Cooperrider, 1995), all published by Ohio State University Press. Another volume of this group is currently being written by John J. Furlow (in prep.).

In addition to the production of these books, the Ohio Flora Project has served as a stimulus for scores of local, county, and regional floristic surveys within the state, as well as a number of studies of individual plant families in Ohio (Cooperrider, 1984, 1995; Furlow, in prep.; Roberts and Stuckey, 1974). These efforts have led to the rapid growth of herbaria at several Ohio museums, colleges, and universities (Cusick and Snider, 1982). Also during this period, Clara G. Weishaupt (1971) published the third edition of a manual that provided identification keys to the state's vascular plants.

In the 1970s, a project organized by Charles C. King and the Ohio Biological Survey resulted in a study of Ohio's endangered and threatened plant species. Edited by Cooperrider (1982), the book includes individual sections on major taxonomic groups. The authors were William Adams, Cooperrider, Marvin L. Roberts, Ray E. Showman, Jerry A. Snider, and Ronald L. Stuckey. Earlier, Stuckey and Roberts (1977) had published a report on the state's rare and endangered aquatic vascular plants. These projects, like those before them, served as incentives for still more research on the nature and composition of the Ohio flora.

Also initiated in the 1970s, and also focused on the state's rare and endangered species, was a long-term effort, the Ohio Natural Heritage Program (Stuckey, 1982; Moseley, 1990). The Heritage Program, established jointly by the Ohio Chapter of The Nature Conservancy and the Ohio Department of Natural Resources (ODNR), continues today, now as a part of the ODNR's Division of Natural Areas and Preserves (DNAP). The fieldwork and research of the DNAP staff over the past twenty years have produced further additions to the state's known native flora and led to the discovery of many new stations for rare Ohio plants. A summary of information about their findings along with accounts of new plant discoveries made by other botanists throughout the state, for example, the 1998 records reported by James S. McCormac (1999), appears annually in the DNAP newsletter.

The DNAP also administers the Ohio Endangered Plant Law, which became effective in 1978 (Moseley, 1990). In that capacity the DNAP maintains an authoritative list, updated biennially, of the plants protected by the law. The final list of the 1990 decade (Ohio Division of Natural Areas and Preserves, 1998), assembled under the direction of Patricia D. Jones and the Ohio Rare Plants Advisory Committee, includes 608 taxa of rare, native vascular plants.

Invading weeds and escapes from cultivation continue to enter the state's flora, some appearing only briefly, others sooner or later becoming established. Because of their potentially deleterious impact on farms, gardens, and nature preserves and other natural areas, these alien species have lately received increased attention. Michael A. Vincent and Allison W. Cusick (1998) recently published an extensive list of alien plants newly discovered in Ohio.

The catalog that follows lists 2,716 species of vascular plants. Of these, 1,785 are regarded as native to the state; this figure includes a few species with both native and nonnative elements. The remaining 931 species are regarded as completely alien in the flora, 507 naturalized and marked with an asterisk, 424 not naturalized and marked with a dagger sign. The catalog also includes 139 interspecific hybrids, 115 of which are naturally occurring hybrids native to Ohio. See Appendix 1 for additional statistics. A list of 132 deleted taxa appears in Appendix 2.

Acknowledgments and Closing Comments

I have been fortunate in having a group of talented coauthors and coeditors with whom to work. I appreciate the effort each of them has devoted to the production of this catalog, studying the flora anew and consulting recent as well as older taxonomic literature.

I thank especially Linda L. Matz, who handled with great skill the complex and difficult word processing involved in preparing the manuscript. The editors, authors, and all those who use the catalog are in her debt.

The authors extend thanks to Jerry M. Baskin and Michael A. Vincent, both of whom read the entire manuscript and made helpful suggestions for its improvement. They thank also George W. Argus, Brian J. Armitage, Harvey E. Ballard Jr., James K. Bissell, Norlyn L. Bodkin, Charles T. Bryson, Veda M. Cafazzo, Philip D. Cantino, J. Richard Carter, Carl F. Chuey, Beverly W. Danielson, Sarah M. Emery, T. Richard Fisher, Richard L. Hauke, Robert R. Haynes, Patricia D. Jones, Charles C. King, Jeffrey D. Knoop, James S. McCormac, Richard S. Mitchell, Sergei L. Mosyakin, Juliana C. Mulroy, Robert F. C. Naczi, Michael J. Oldham, Richard K. Rabeler, Anton A. Reznicek, Gregory J. Schneider, Susan L. Sherman-Broyles, S. Galen Smith, Jerry A. Snider, Victor G. Soukup, Ronald L. Stuckey, Sue A. Thompson, Steven T. Trimble, Gordon C. Tucker, Michael A. Vincent, Edward G. Voss, Warren H. Wagner Jr., Alan T. Whittemore, George J. Wilder, Hugh D. Wilson, and Michael D. Windham for aid with various specific problems. The authors of the monocot section thank in particular Mary E. Barkworth for guidance on some problems in the Poaceae; the treatment of that family here is based in part on Barkworth, Capels, and Vorobik (in prep.).

For their work in converting the manuscript into a book, we are indebted to the excellent staff of the Ohio State University Press. We thank former chief Guy L. Denny and current chief W. Stuart Lewis of the ODNR Division of Natural Areas and Preserves and members of the Ohio Natural Areas Council for assistance in obtaining financial support for the book's publication.

I am grateful to Miwako K. Cooperrider for help in many ways, to Julie A. Cooperrider for help in preparing the indexes, and to John A. Cooperrider for providing the cover sketch. I thank Guy Denny for inviting me to serve as second author of the next section, "Natural History of the Ohio Flora."

■

The idea of preparing a *Seventh Catalog* first occurred to me in 1960. My Kent State colleague J. Arthur Herrick loaned me a copy of Schaffner's 1932 *Catalog* in which his brother, Ervin M. Herrick, had noted species distribution records for several counties of northeastern Ohio. I used the records to construct a provisional checklist to guide my field work and quickly realized the value that a new catalog would have. Preliminary lists of various Ohio plant groups with statements of county-by-county distribution had been prepared for, and distributed by, the Ohio Flora Committee. They included those by William Adams for the pteridophytes, E. Lucy Braun the monocots exclusive of grasses, William G. Gambill the legumes, Robert W. Long the composites, and Clara G. Weishaupt the grasses (Cooperrider, 1961). These lists, as well as the OSU Press books that were to follow, would provide a great amount of the basic information needed for such an effort.

A decade later, Weishaupt's 1971 *Manual*, noted earlier, provided an updated accounting of the state's vascular plants. As new information accumulated in the 1970s and 1980s, I drafted a preliminary checklist and began planning a new catalog. In 1978, Ohio State botanist Ronald L. Stuckey sent me a copy of a talk he had given that included a list of several botanical references needed for Ohio, one of them a new checklist of the state's flora.

In 1994, John Kartesz, who had prepared his own preliminary checklist for Ohio, suggested that we combine our efforts and invite Allison Cusick to join us. Barbara Andreas, John Freudenstein, and John Furlow agreed to cover plant groups in which they had special expertise, and Guy Denny agreed to adapt his essay on Ohio's natural history for its use here. This book is the result of that collaboration.

The cutoff date for adding new information to the catalog was December 31, 1999. Work on the Ohio flora continues, and the new century will bring its own set of additions and revisions.

For Ohio plants, the present in particular is a time of considerable taxonomic change, stemming in large part from research for the Flora of North America Project (Flora of North America Editorial Committee, 1993a). That project, and the nomenclatural and classificatory innovations engendered by it, will likely continue for the next two or three decades — affecting the scientific names of various groups of Ohio plants. For some taxa in the following list, the choice of a scientific name today is a matter of judgment. In a few such instances and also in some procedural details I have made or accepted decisions regarding entries in the catalog with which a coworker on the project has disagreed. Any errors that exist are my responsibility.

The editors and authors join me in the hope that the *Catalog* will prove of value to all whose interest or work leads them to deal with Ohio's vascular plants.

NATURAL HISTORY OF THE OHIO FLORA

Guy L. Denny and Tom S. Cooperrider

What is now the state of Ohio was once a living tapestry of several major plant communities, including prairies, bogs, fens, oak openings, marshes, and sedge meadows, set within a vast matrix of diverse forest types. Robert B. Gordon's (1966) map of Ohio's natural vegetation at the time of the earliest land surveys provides the best image available of the Ohio wilderness before European settlement. The map and associated book (Gordon, 1969), published as a cooperative project of the Ohio Biological Survey and the Natural Resources Institute of The Ohio State University, were based largely on field notes, journals, and plot maps of the earliest land surveyors, who explored the region in the late eighteenth and early nineteenth centuries.

Although most of Ohio's natural landscape has since been greatly altered by agricultural, industrial, and residential development, an impressive and diverse set of natural communities remains—albeit greatly fragmented and reduced in extent. Many of these remnant communities are in nature preserves, scattered among the state's eighty-eight counties (see map, p. x).

To understand and appreciate the state's rich natural diversity, we need to examine a series of past geological and ecological events that helped mold the vegetation of present-day Ohio. Our study begins near the close of the Tertiary Period, about two to two and a half million years before the present. At that time, neither the Great Lakes nor the Ohio River had yet come into existence. However, there was in existence an impressive river system known as the Teays that traversed and influenced much of what was to become Ohio.

The Teays River Influence

The headwaters of the Teays were in the Appalachian Mountains of North Carolina. From there the great river flowed in a northwesterly direction entering southern Ohio at the edge of the Appalachian Plateau immediately east of the present city of Portsmouth. It flowed north to central Ohio, then veered northwestward,

Adapted from Ohio Division of Natural Areas and Preserves (1996).

exiting the state near Grand Lake Saint Marys. From Ohio it continued west across Indiana and Illinois, eventually in western Illinois joining the ancestral Mississippi River. Over the millions of years of its existence, the Teays provided a continuum of riverine habitats that served as a corridor for the northerly range extension of Appalachian plants and animals.

In southern Ohio, within isolated protected pockets along the now abandoned ancient channels of the Teays River, some of these Appalachian Mountain species flourish today, remnants of the Teays influence in Ohio. Among them are a number of flowering trees and shrubs, including great rhododendron (*Rhododendron maximum*), flame azalea (*Rhododendron calendulaceum*), Canby's mountain-lover (*Paxistima canbyi*), umbrella magnolia (*Magnolia tripetala*), and bigleaf magnolia (*Magnolia macrophylla*). Rhododendron Cove Preserve in Hocking County supports the largest population of great rhododendron in Ohio. The best displays of the two magnolias can be seen at Lake Katharine Preserve in Jackson County.

The Boreal Forest Influence

About two million years ago, the climate of North America became cooler than at present, giving rise to several major continental ice sheets, at least three of which advanced across the northern and western two-thirds of Ohio. With the beginning of the Pleistocene or Ice Age, a new chapter began in the story of the origin of Ohio's natural diversity. Where these massive ice sheets covered the landscape, all existing vegetation was destroyed, though some species managed to extend their ranges southward beyond the glacial limit. Along with the ice sheets came a new element of diversity composed of northern plants and animals that preceded the slowly moving glaciers southward. At a few sites, such as in deep bedrock gorges and in bogs and fens, where local conditions have somewhat approximated the cool, moist environment associated with glaciation, some of the boreal plants have persisted in our modern landscape.

After a gradual, long-term warming trend, the Wisconsinan Glacier, the last and most recent of the continental ice sheets to invade Ohio, began a retreat northward into eastern Canada about 18,000 years ago, finally leaving the state about 12,000 years ago. As the glacier melted, huge blocks of ice often broke off and were buried in thick glacial deposits of clay, silt, sand, and gravel. Eventually, when the ice blocks melted, deep kettle lakes were left in their depressions and northern bog plants became established around the shores of these lakes. The plants include rare orchids such as grass-pink (*Calopogon tuberosus*) and rose pogonia (*Pogonia ophioglossoides*), wild cranberries (*Vaccinium macrocarpon* and *V. oxycoccos*), and carnivorous plants such as round-leaved sundew (*Drosera rotundifolia*) and pitcher-plant (*Sarracenia purpurea*). Because of localized, special environmental conditions, many of these Ice Age sphagnum peat bogs survive today. Some of the best state nature preserves for viewing sphagnum peat bog communities include Cooperrider-Kent Bog and Triangle Lake Bog in Portage County, and Cranberry Bog in Licking County.

In addition to these acid sphagnum peat bogs, there is another type of Ice Age peatland community that occurs in Ohio. These are alkaline or calcareous bogs, now more commonly referred to as fens. Boreal plants as well as Atlantic Coastal Plain species have been able to survive in Ohio fens, where cold, alkaline springs saturate the soil, creating conditions not unlike those that probably occurred along the shores of ancient glacial lakes. Here are found such species as shrubby cinquefoil (*Potentilla fruticosa*), grass-of-parnassus (*Parnassia glauca*), Ohio goldenrod (*Solidago ohioensis*), showy lady's-slipper (*Cypripedium reginae*), and fringed gentian (*Gentianopsis procera*). These calcareous ecosystems have the highest concentrations of state-listed rare species of any plant community in the state. Examples are Cedar Bog Preserve in Champaign County, Jackson Bog Preserve in Stark County, and Frame Lake/Herrick Fen Preserve in Portage County. In addition to boreal and Atlantic Coastal Plain species, many fens, especially those in west-central Ohio, have an abundance of prairie plants such as queen-of-the-prairie (*Filipendula rubra*), big bluestem grass (*Andropogon gerardii*), prairie dock (*Silphium terebinthinaceum*), and spiked blazing-star (*Liatris spicata*). Good examples of this type of habitat include Prairie Road Fen Preserve in Clark County, Zimmerman Prairie Preserve in Greene County, and Springfield/Gallagher Fen Preserve in Champaign County.

The Oak Openings

Located in northwestern Ohio is an extensive expanse of dry sandy ridges and knolls known as the Oak Openings. The area covers more than 150 square miles, encompassing western Lucas County as well as portions of Henry and Fulton Counties, and is defined by its conspicuous soils of fine yellowish sand. The name is derived from the open stands of stunted oaks that survive on the sandy fields.

About 14,000 years ago, toward the close of the Ice Age, as the ice lobes melted back into the preglacial Great Lakes watershed, tremendous volumes of meltwater were ponded between the glacier and the extensive Fort Wayne end moraine. As the ice edge shifted back and forth, lake levels in the Erie Basin fluctuated from time to time, producing a succession of more than ten distinctive glacial lakes, each of which has been named. The highest in elevation was Lake Maumee, which stood over 225 feet above the present level of Lake Erie. Each different lake level is marked today by its own series of beaches and sand bars.

The Oak Openings region was formed from the wind-blown beach sand of glacial Lake Warren, which existed about 12,000 years ago. The yellowish-brown sand was washed into Lake Warren from Michigan by the ancestral Detroit River. Once the lake level dropped, as the Niagara outlet was freed of glacial ice, these beaches were left high and dry to be reshaped by prevailing winds.

Where the sand is thick, subsurface drainage is excessive and the soils are dry and acid. Where the sand is thin, ground water trapped by the underlying impervious lake clay creates shallow ponds, wet meadows, and swamp forests. As a result of the abnormal soil conditions, the Oak Openings region is characterized by an unusually large number of rare plants including both prairie species and Atlantic Coastal Plain species. Irwin Prairie, Kitty Todd, and Lou Campbell Preserves, all situated in western Lucas County, are excellent examples of the Oak Openings region.

The Great Black Swamp

Once the Niagara outlet was freed of glacial ice, not only were the sand deposits north of the Maumee River left behind, but south of the river, the poorly drained glacial lake bed was exposed. Owing to its flatness and impervious clay liner, this land remained flooded for most of the year, giving rise to an extensive swamp forest dominated by American elm (*Ulmus americana*), black ash (*Fraxinus nigra*), red maple (*Acer rubrum*), pin oak (*Quercus palustris*), and swamp white oak (*Q. bicolor*). Running from Lake Erie to Indiana, the swamp was about a hundred miles long and twenty to forty miles wide. This dense, mosquito-infested morass, officially named the Great Black Swamp in 1812, was a significant impediment to travel and consequently settlement of northwestern Ohio. In 1859 a law providing for public ditches in the area set the stage for the demise of the Great Black Swamp. By the end of the nineteenth century, nearly the entire area, Ohio's last great wilderness, had been drained and converted to farm land. Goll Woods State Nature Preserve supports an old-growth swamp forest characteristic of all that remains of the Great Black Swamp.

The Prairie Influence

Pollen profiles taken from sphagnum peat bogs in the Midwest reveal that about 4,000 – 8,000 years ago, North America went through an extended period of warming and drought-like conditions that favored expansion of the western tallgrass prairie eastward into the deciduous forest region. During this Xerothermic period, now more commonly referred to as the Hypsithermal Interval, extensive areas of prairie became a part of the Ohio landscape. A region of continuous prairie, called the Prairie Peninsula, extended eastward across the Mississippi River, through Illinois and Indiana, and into Ohio.

Early settlers described the prairies of Ohio as a sea of tall grasses and colorful wildflowers. The grasses were interrupted only by numerous scattered groves of oaks and hickories, especially bur oaks (*Quercus macrocarpa*), which were particularly resistant to the raging prairie fires that often occurred in the fall and early spring. The tall grasses were primarily big bluestem (*Andropogon gerardii*)—also known as turkey-foot grass, Indian grass (*Sorghastrum nutans*), and prairie cord grass (*Spartina pectinata*).

As the climate changed toward one favoring deciduous forest communities, Ohio's portion of the Prairie Peninsula became fragmented. Writing on the history of Ohio's prairies, K. Roger Troutman reports (in Lafferty, 1979) that during pioneer times, there were about 1,000 square miles of prairie in Ohio, divided among more than 300 individual prairie openings. The openings ranged in size from a few acres to several thousand acres. Most of these sites, with their ancient deep, rich soils, now support farm crops rather than prairie communities. Nevertheless, small pockets of prairie still occur and are well represented in several state nature preserves. Some of the best examples of tallgrass prairie preserves are Bigelow Cemetery Prairie and Smith Cemetery Prairie in Madison County, Milford Center Prairie in Union County, and Compass Plant Prairie in Lawrence County.

A separate group of prairies, those of the unglaciated Blue Grass Region of Adams County in southwestern Ohio, are distinctively different from their tallgrass counterparts in glaciated Ohio. It is believed that they may predate the Wisconsinan glaciation. Even though they share some of the same species with other Ohio prairies, a number of the plants that grow there, such as spider milkweed (*Asclepias viridis*) and bluehearts (*Buchnera americana*), are more typical of the Cedar Glades of the Missouri Ozarks. Lynx Prairie, Adams Lake Prairie, and nearby Chaparral Prairie Preserves are excellent examples of these communities.

The Shoreline of Lake Erie

The extensive marshes of western Lake Erie are one of the natural treasures of Ohio, supporting a rich diversity of both plant and animal life. Sheldon Marsh Preserve in Erie County, Magee Marsh State Wildlife Area in Ottawa County, and Ottawa National Wildlife Refuge in Ottawa and Lucas Counties are good examples of western Lake Erie marsh communities. By comparison, one of the best examples of an inland marsh community can be seen at Springville Marsh Preserve in Seneca County.

The shores of eastern Lake Erie, on the other hand, have a series of sandy beaches that are home to another element of natural diversity in Ohio, representatives of the Atlantic shoreline flora. The migration of Atlantic Coastal Plain species into the Great Lakes Region probably took place thousands of years ago during Algonquin time, a period when the sea extended into the present basin of Lake Ontario owing to a depression of the bedrock basin brought about by the weight of continental glaciation.

Although subsequent crustal rebound of the bedrock has since isolated these plants from the sea, they persist on the few relatively undisturbed sandy beaches in northeastern Ohio as well as elsewhere along the Great Lakes. Two of the best sites for viewing Atlantic Coastal Plain species such as American beach grass (*Ammophila breviliqulata*), inland beach pea (*Lathyrus japonicus*), and inland sea rocket (*Cakile edentula*), are Headlands Dunes Preserve in Lake County and Ashtabula Beach in Ashtabula County.

The Northern Allegheny Mountain Influence

Another element of natural diversity is the Northern Allegheny Mountain influence occurring in the northeastern corner of the state, especially in Ashtabula

County. Just as the Teays River influenced Appalachian Mountain vegetation in southern Ohio, so too its contemporary, the Pittsburgh River, with headwaters in the Allegheny Mountains of Pennsylvania, influenced the vegetation of extreme northeastern Ohio.

The special plant communities of northeastern Ohio are probably the result of the high amount of snowfall in the region, the acid soil continuum into Pennsylvania, the preglacial watershed affiliation with the Allegheny Mountains, or a combination of all three. Certainly as the last glacier retreated from this part of Ohio, the newly exposed rolling landscape was well suited for the invasion of adjacent Allegheny Mountain species, moving westward into northeastern Ohio. Some believe this range expansion is still occurring.

Hemlock *(Tsuga canadensis)*, yellow birch *(Betula alleghaniensis)*, gray birch *(Betula populifolia)*, and even white pine *(Pinus strobus)* are significant components of poorly drained woodlands of this region. These wet woodlands, in turn, support a variety of northern Allegheny species such as hobblebush *(Viburnum alnifolium)*, bluebead-lily *(Clintonia borealis)*, painted trillium *(Trillium undulatum)*, starflower *(Trientalis borealis)*, Carolina spring-beauty *(Claytonia caroliniana)*, robin-run-away *(Dalibarda repens)*, and rose twisted-stalk *(Streptopus roseus)*. Pallister Nature Preserve in Ashtabula County is an example of this type of habitat.

The Old-Growth Forest

After the Ice Age ended, deciduous forests became established over most of Ohio. In some areas, the subsequent period of forestation was interrupted for a time by the Prairie Peninsula, but as cooler, moister climatic conditions returned a few thousand years ago, a large part of the grassland in Ohio also, once again, gave way to deciduous forests.

As a result, when the first European settlers reached the Ohio country, they encountered one of the most magnificent old-growth forests on the face of the earth. Their journals describe huge trees six feet and more in diameter, towering sixty to eighty feet before the first limbs. Although certainly many of the settlers appreciated the grandeur of the primeval forest, most viewed the dense woods as an impediment to their agricultural lifestyle. The giant trees had to be removed before the rich forest soils could be tilled and planted. The removal of forests began in earnest by the late 1700s, when Ohio started to forge its place as a major agricultural and industrial power.

Even though Ohio has more forest cover now than it did a hundred years ago, it is primarily young growth replacement forest cover. There are two major forest types in Ohio today: (1) Beech Forests—characterized by a large fraction of beech *(Fagus grandifolia)*, sugar maple *(Acer saccharum)*, red oak *(Quercus rubra)*, white ash *(Fraxinus americana)*, and white oak *(Quercus alba)*— occur primarily in western, glaciated Ohio. Beech forests occur also in the glaciated Allegheny Plateau

Region of northeastern Ohio, where they are generally associated with tulip-tree (*Liriodendron tulipifera*), red maple (*Acer rubrum*), and/or sugar maple. (2) Mixed Oak Forests—characterized by a large fraction of white oak, black oak (*Quercus velutina*), and hickories (*Carya* spp.)—tend to be most common in east-central and southeastern, unglaciated Ohio.

The primeval forests, which once covered most of the state, have been reduced to little more than a few scattered stands. Several fine forest remnants are represented within the state nature preserve system. Examples include Hueston Woods in southwestern Ohio, Goll Woods in northwestern Ohio, Johnson Woods in north-central Ohio, and Fowler Woods in central Ohio. Many of these woodland preserves have spectacular displays of spring wildflowers in April and May.

■

Readers are referred to Gordon (1966, 1969) and Lafferty (1979) for additional information on Ohio's natural heritage, and to Ohio Division of Natural Areas and Preserves (1996) for additional information on the nature preserves listed here.

CATALOG OF VASCULAR PLANTS

The nomenclature, circumscription, and sequence of phyla (= divisions), classes, subclasses, orders, and families are based on Cronquist (1981), Gleason and Cronquist (1991), and/or Flora of North America Editorial Committee (1993b, 1997). The nomenclature and circumscription of taxa below the rank of family, i.e., genera, species, subspecies, varieties, forms, and interspecific hybrids, are based on floristic and taxonomic research publications from a variety of sources; the names of these taxa are presented in alphabetical sequence.

Scientific names of the species, hybrids, and infraspecific taxa in the flora are printed in **boldface**. Names preceded by an asterisk or dagger sign are of taxa regarded as alien in the Ohio flora. An asterisk (*) designates plants that are naturalized, that is, ones established and maintaining themselves, either locally or more widely in the state. A dagger sign (†) designates either plants that are adventive or plants that have spread only into the immediate vicinity of a site of cultivation, that is, ones that are not established and are not maintaining themselves in the flora. The distinction between the two classes of aliens is not always clearcut; the decisions adopted here represent our best knowledge at the present time. All other names, those lacking a symbol, are of taxa regarded as native (indigenous) in all or some part of their Ohio range. This last group includes plants such as **Campsis radicans,** TRUMPET-CREEPER or TRUMPET-VINE, native in southern Ohio, but naturalized at scattered sites in northern counties.

The authors (authorities) of scientific names are mostly those ascribed by Kartesz (1994). The forms or abbreviations of authors' names are those recommended by Brummitt and Powell (1992).

Following the style of our major regional manual, Gleason and Cronquist (1991), the rank of variety is used in preference to that of subspecies whenever possible. For groups that have been given names at both ranks, the subspecific epithet is listed in *italics*. In some cases the varietal and subspecific epithets are the same: e.g., **Equisetum hyemale var. affine** (subsp. *affine*); in others they are different: e.g., **Acer saccharum var. viride** (subsp. *nigrum*). In most instances in which the only variety occurring in Ohio is the type variety, no varietal name is given, e.g., **Linum sulcatum** instead of **Linum sulcatum var. sulcatum,** because of the sometimes uncertain validity of the other varieties. Exceptions to this practice are made only when needed for clarity. At the rank of forma, only the most noteworthy groups are included.

Interspecific hybrids are listed by formula, e.g., **Lysimachia quadrifolia** × **L. terrestris**, preceded by a name if a hybrid name is available, e.g., **Lysimachia** × **producta**. Hybrids are arranged alphabetically at the end of the species list for their genus. No mention is made of hybrids between different varieties of the same species.

Common or vernacular names are printed in Capitals and small capitals. They are taken primarily from the works listed in the next paragraph. They also include common names used in *Hortus Third* (Bailey Hortorium Staff, 1976) and those used in recent publications of the Ohio Department of Natural Resources, the Ohio Biological Survey, and the Ohio Chapter of The Nature Conservancy.

In *italics* are alternate scientific names for these plants, ones not used here but used in one or more of the following works: Braun (1961, 1967), Weishaupt (1971), Fisher (1988), Gleason and Cronquist (1991), Cooperrider (1995), Flora of North America Editorial Committee (1993b, 1997), and Furlow (1997). Kartesz (1994) and Kartesz and Meacham (1999) have more extensive lists of synonyms.

—T.S.C.

Pteridophytes

Allison W. Cusick

Phylum **LYCOPODIOPHYTA.** CLUB-MOSSES, SPIKE-MOSSES, QUILLWORTS

Class **Lycopodiopsida**

Order **Lycopodiales**

- LYCOPODIACEAE. CLUB-MOSS FAMILY

Diphasiastrum digitatum (Dill. ex A. Braun) Holub GROUND-PINE, SOUTHERN RUNNING-PINE. (Incl. *Lycopodium complanatum* L. var. *flabelliforme* Fernald, *L. flabelliforme* (Fernald) Blanch.; *L. digitatum* Dill. ex A. Braun)

Diphasiastrum tristachyum (Pursh) Holub BLUE GROUND-PINE, BLUE GROUND-CEDAR. (*Lycopodium tristachyum* Pursh)

Diphasiastrum × habereri (House) Holub **(Diphasiastrum digitatum × D. tristachyum)** HABERER'S GROUND-PINE. (*Lycopodium × habereri* House)

Huperzia lucidula (Michx.) Trevis. SHINING CLUB-MOSS, SHINING FIR-MOSS. (*Lycopodium lucidulum* Michx.)

Huperzia porophila (F. E. Lloyd & Underw.) Holub ROCK CLUB-MOSS, ROCK FIR-MOSS. (*Lycopodium porophilum* F. E. Lloyd & Underw.)

Huperzia × bartleyi (Cusick) Kartesz & Gandhi **(Huperzia lucidula × H. porophila)** BARTLEY'S CLUB-MOSS. (*Lycopodium × bartleyi* Cusick)

Lycopodiella inundata (L.) Holub NORTHERN BOG CLUB-MOSS. (*Lycopodium inundatum* L.)

Lycopodiella margueritae J. G. Bruce, W. H. Wagner, & Beitel NORTHERN PROSTRATE CLUB-MOSS.

Lycopodiella subappressa J. G. Bruce, W. H. Wagner, & Beitel NORTHERN APPRESSED CLUB-MOSS.

Lycopodium clavatum L. RUNNING CLUB-MOSS, COMMON CLUB-MOSS.

Lycopodium dendroideum Michx. PRICKLY TREE CLUB-MOSS. (*L. obscurum* L. var. *dendroideum* (Michx.) D. C. Eaton)

Lycopodium hickeyi W. H. Wagner, Beitel, & R. C. Moran ROUND-BRANCHED TREE CLUB-MOSS, HICKEY'S TREE CLUB-MOSS. (*L. obscurum* L. var. *isophyllum* Hickey)

Lycopodium lagopus (Laest. ex C. Hartm.) G. Zinserl. ex Kuzen. ONE-CONED CLUB-MOSS, RABBIT'S-FOOT CLUB-MOSS. (*L. clavatum* L. var. *monostachyon* Grev. & Hook.)

Lycopodium obscurum L. FLAT-BRANCHED TREE CLUB-MOSS.

Class **Isoetopsida**

Order **Selaginellales**

- SELAGINELLACEAE. SPIKE-MOSS FAMILY

Selaginella apoda (L.) Spring MEADOW SPIKE-MOSS.

Selaginella eclipes W. R. Buck MIDWEST SPIKE-MOSS, HIDDEN SPIKE-MOSS.

Selaginella rupestris (L.) Spring ROCK SPIKE-MOSS, DWARF SPIKE-MOSS.

Order **Isoetales**

- ISOETACEAE. QUILLWORT FAMILY

Isoetes echinospora Durieu SPINY-SPORED QUILLWORT.

Isoetes engelmannii A. Braun Appalachian Quillwort, Engelmann's Quillwort.

Phylum EQUISETOPHYTA. HORSETAILS, SCOURING-RUSHES

Class Equisetopsida

Order Equisetales

■ EQUISETACEAE. Horsetail Family

Equisetum arvense L. Field Horsetail.

Equisetum fluviatile L. River Horsetail, Pipes.

Equisetum hyemale L. var. affine (Engelm.) A. A. Eaton Tall Scouring-rush. (subsp. affine (Engelm.) Calder & Roy L. Taylor)

Equisetum laevigatum A. Braun Smooth Scouring-rush.

Equisetum sylvaticum L. Woodland Horsetail.

Equisetum variegatum Schleich. ex F. Weber & D. Mohr Variegated Scouring-rush.

Equisetum × ferrissii Clute (Equisetum hyemale × E. laevigatum) Ferriss' Scouring-rush.

Equisetum × mackaii (Newman) Brichan (Equisetum hyemale × E. variegatum) Mackay's Scouring-rush.

Equisetum × nelsonii (A. A. Eaton) J. H. Schaffn. (Equisetum laevigatum × E. variegatum) Nelson's Scouring-rush.

Phylum POLYPODIOPHYTA. FERNS

Class Polypodiopsida

Order Ophioglossales

■ OPHIOGLOSSACEAE. Adder's-tongue Family

Botrychium biternatum (Savigny) Underw. Sparse-lobed Grape Fern.

Botrychium dissectum Spreng. Dissected Grape Fern. (Incl. B. obliquum Muhl. ex Willd.)

Botrychium lanceolatum (S. G. Gmel.) Ångstr. var. angustisegmentum Pease & A. H. Moore Triangle Grape Fern, Narrow Triangle Moonwort. (subsp. angustisegmentum (Pease & A. H. Moore) R. T. Clausen)

Botrychium matricariifolium (Döll) A. Braun ex W. D. J. Koch Daisy-leaved Grape Fern, Daisy-leaved Moonwort.

Botrychium multifidum (S. G. Gmel.) Rupr. Leathery Grape Fern.

Botrychium oneidense (Gilbert) House Blunt-lobed Grape Fern.

Botrychium simplex E. Hitchc. Least Grape Fern, Least Moonwort.

Botrychium virginianum (L.) Sw. Rattlesnake Fern.

Ophioglossum engelmannii Prantl Limestone Adder's-tongue.

Ophioglossum pusillum Raf. Northern Adder's-tongue. (O. vulgatum var. pseudopodum (S. F. Blake) Farw.)

Ophioglossum vulgatum L. Southern Adder's-tongue. (Incl. var. pycnostichum Fernald)

Order Polypodiales

■ OSMUNDACEAE. Royal Fern Family

Osmunda cinnamomea L. Cinnamon Fern.
Osmunda claytoniana L. Interrupted Fern.
Osmunda regalis L. var. spectabilis (Willd.) A. Gray Royal Fern.

■ LYGODIACEAE. Climbing Fern Family

Lygodium palmatum (Bernh.) Sw. American Climbing Fern, Hartford Fern.

■ PTERIDACEAE. Maidenhair Fern Family

Adiantum pedatum L. Northern Maidenhair, Maidenhair Fern.

Pellaea atropurpurea (L.) Link Purple Cliff-brake.

Pellaea glabella Mett. ex Kuhn Smooth Cliff-brake.

■ VITTARIACEAE. Shoestring Fern Family

Vittaria appalachiana Farrar & Mickel Appalachian Shoestring Fern, Appalachian

GAMETOPHYTE. (Treated by some authors as a part of *V. lineata* (L.) Sm.)

■ HYMENOPHYLLACEAE. FILMY FERN FAMILY

Trichomanes boschianum J. W. Sturm APPALACHIAN FILMY FERN.

Trichomanes intricatum Farrar WEFT FERN, APPALACHIAN TRICHOMANES.

■ DENNSTAEDTIACEAE. HAY-SCENTED FERN FAMILY

Dennstaedtia punctilobula (Michx.) T. Moore HAY-SCENTED FERN.

Pteridium aquilinum (L.) Kuhn BRACKEN.
var. latiusculum (Desv.) Underw. ex A. Heller EASTERN BRACKEN.
var. pseudocaudatum (Clute) A. Heller TAILED BRACKEN.

■ THELYPTERIDACEAE. MARSH FERN FAMILY

Phegopteris connectilis (Michx.) Watt LONG BEECH FERN, NORTHERN BEECH FERN. (*Thelypteris phegopteris* (L.) Sloss.)

Phegopteris hexagonoptera (Michx.) Fée BROAD BEECH FERN, SOUTHERN BEECH FERN. (*Thelypteris hexagonoptera* (Michx.) Weath.)

Thelypteris noveboracensis (L.) Nieuwl. NEW YORK FERN.

Thelypteris palustris Schott **var. pubescens** (G. Lawson) Fernald MARSH FERN.

■ BLECHNACEAE. CHAIN FERN FAMILY

Woodwardia areolata (L.) T. Moore NETTED CHAIN FERN.

Woodwardia virginica (L.) Sm. VIRGINIA CHAIN FERN.

■ ASPLENIACEAE. SPLEENWORT FAMILY

Asplenium bradleyi D. C. Eaton BRADLEY'S SPLEENWORT.

Asplenium montanum Willd. MOUNTAIN SPLEENWORT.

Asplenium pinnatifidum Nutt. LOBED SPLEENWORT.

Asplenium platyneuron (L.) Britton, Sterns, & Poggenb. EBONY SPLEENWORT.

Asplenium resiliens Kunze BLACK-STEMMED SPLEENWORT.

Asplenium rhizophyllum L. WALKING FERN. (*Camptosorus rhizophyllus* (L.) Link)

Asplenium ruta-muraria L. WALL-RUE.

Asplenium trichomanes L. MAIDENHAIR SPLEENWORT.
subsp. quadrivalens D. E. Mey.
subsp. trichomanes

Asplenium × clermontiae Syme (**Asplenium ruta-muraria × A. trichomanes**) CLERMONT SPLEENWORT.

Asplenium × ebenoides R. R. Scott (**Asplenium platyneuron × A. rhizophyllum**) SCOTT'S SPLEENWORT.

Asplenium × gravesii Maxon (**Asplenium bradleyi × A. pinnatifidum**) GRAVES'S SPLEENWORT.

Asplenium × inexpectatum (E. L. Braun ex Friesner) C. V. Morton (**Asplenium rhizophyllum × A. ruta-muraria**) UNEXPECTED SPLEENWORT.

Asplenium × kentuckiense T. N. McCoy (**Asplenium pinnatifidum × A. platyneuron**) KENTUCKY SPLEENWORT.

Asplenium × trudellii Wherry (**Asplenium montanum × A. pinnatifidum**) TRUDELL'S SPLEENWORT.

■ DRYOPTERIDACEAE. WOOD FERN FAMILY

Athyrium filix-femina (L.) Roth ex Mert. LADY FERN.
var. angustum (Willd.) G. Lawson NORTHERN LADY FERN. (subsp. *angustum* (Willd.) R. T. Clausen; incl. var. *michauxii* (Spreng.) Farw.)
var. asplenioides (Michx.) Farw. SOUTHERN LADY FERN. (subsp. *asplenioides* (Michx.) Hultén)

†**Cyrtomium falcatum** (L. f.) C. Presl   Holly Fern.

Cystopteris bulbifera (L.) Bernh.   Bulblet Bladder Fern.

Cystopteris fragilis (L.) Bernh. **var. fragilis**   Brittle Fern, Fragile Fern.

Cystopteris protrusa (Weath.) Blasdell   Lowland Fragile Fern, Lowland Brittle Fern.

Cystopteris tennesseensis Shaver   Tennessee Bladder Fern.

Cystopteris tenuis (Michx.) Desv.   Cliff Fragile Fern, Mackay's Brittle Fern. (*C. fragilis* var. *mackayi* G. Lawson)

Cystopteris × illinoensis R. C. Moran **(Cystopteris bulbifera × C. tenuis)**   Illinois Fragile Fern.

Cystopteris × wagneri R. C. Moran **(Cystopteris tennesseensis × C. tenuis)**   Wagner's Fragile Fern.

Cystopteris protrusa × C. tenuis

Deparia acrostichoides (Sw.) M. Kato   Silvery Glade Fern, Silvery Spleenwort. (*Athyrium thelypterioides* (Michx.) Desv., *Diplazium acrostichoides* (Sw.) Butters)

Diplazium pycnocarpon (Spreng.) M. Broun   Narrow-leaved Glade Fern. (*Athyrium pycnocarpon* (Spreng.) Tidestr.)

Dryopteris carthusiana (Vill.) H. P. Fuchs   Spinulose Wood Fern, Spinulose Shield Fern. (Incl. *D. spinulosa* (O. F. Müll.) Watt, *D. austriaca* (Jacq.) Woyn. ex Schinz & Thell. var. *spinulosa* (O. F. Müll.) Fisch.)

Dryopteris clintoniana (D. C. Eaton) Dowell   Clinton's Wood Fern. (*D. cristata* var. *clintoniana* (D. C. Eaton) Underw.)

Dryopteris cristata (L.) A. Gray   Crested Wood Fern, Swamp Shield Fern.

Dryopteris goldiana (Hook. ex Goldie) A. Gray   Goldie's Fern, Giant Wood Fern.

Dryopteris intermedia (Muhl. ex Willd.) A. Gray   Evergreen Wood Fern, Fancy Fern. (*D. austriaca* (Jacq.) Woyn. ex Schinz & Thell. var. *intermedia* (Muhl. ex Willd.) C. V. Morton)

Dryopteris marginalis (L.) A. Gray   Leathery Wood Fern, Marginal Wood Fern, Marginal Shield Fern.

Dryopteris × boottii (Tuck.) Underw. **(Dryopteris cristata × D. intermedia)**   Boott's Wood Fern.

Dryopteris × neowherryi W. H. Wagner **(Dryopteris goldiana × D. marginalis)**   Wherry's Wood Fern.

Dryopteris × pittsfordensis Slosson **(Dryopteris carthusiana × D. marginalis)**   Pittsford Wood Fern.

Dryopteris × slossoniae Wherry ex Lellinger **(Dryopteris cristata × D. marginalis)**   Slosson's Wood Fern.

Dryopteris × triploidea Wherry **(Dryopteris carthusiana × D. intermedia)**   Triploid Wood Fern.

Dryopteris × uliginosa (A. Braun ex Dowell) Druce **(Dryopteris carthusiana × D. cristata)**   Marsh Shield Fern.

Gymnocarpium appalachianum Pryer & Haufler   Appalachian Oak Fern.

Gymnocarpium dryopteris (L.) Newman   Common Oak Fern.

Matteuccia struthiopteris (L.) Tod. **var. pensylvanica** (Willd.) C. V. Morton   Ostrich Fern. (*M. pensylvanica* (Willd.) Raymond)

Onoclea sensibilis L.   Sensitive Fern.

Polystichum acrostichoides (Michx.) Schott   Christmas Fern.

Woodsia ilvensis (L.) R. Br.   Rusty Woodsia, Rusty Cliff Fern.

Woodsia obtusa (Spreng.) Torr.   Blunt-lobed Woodsia, Blunt-lobed Cliff Fern.

■ POLYPODIACEAE.   Polypody Family

Pleopeltis polypodioides (L.) E. G. Andrews & Windham **var. michauxiana** (Weath.) E. G. Andrews & Windham   Little Gray Polypody, Resurrection Fern. (*Polypodium polypodioides* (L.) Watt var. *michauxianum* Weath.)

Polypodium appalachianum Haufler & Windham   Appalachian Polypody.

Polypodium virginianum L.   Rock Polypody, Rockcap Fern.

Polypodium appalachianum × P. virginianum

Order **Marsileales**

- MARSILEACEAE.   WATER-CLOVER FAMILY

*Marsilea quadrifolia L.   EURASIAN WATER-
  CLOVER.

Order **Salviniales**

- AZOLLACEAE.   MOSQUITO FERN FAMILY

Azolla caroliniana Willd.   NORTHERN
  MOSQUITO FERN.

Gymnosperms

Allison W. Cusick

Phylum **PINOPHYTA
(CONIFEROPHYTA).** GYMNOSPERMS

Class **Pinopsida**

Order **Pinales**

■ PINACEAE.   PINE FAMILY

† **Abies balsamea** (L.) Mill.   BALSAM FIR.
Larix laricina (Du Roi) K. Koch   TAMARACK.
† **Pinus banksiana** Lamb.   JACK PINE.
Pinus echinata Mill.   SHORTLEAF PINE, YELLOW PINE.
† **Pinus nigra** Arn.   AUSTRIAN PINE.
† **Pinus resinosa** Aiton   RED PINE, NORWAY PINE.
Pinus rigida Mill.   PITCH PINE.
Pinus strobus L.   EASTERN WHITE PINE.
* **Pinus sylvestris** L.   SCOTS PINE, SCOTCH PINE.
Pinus virginiana Mill.   VIRGINIA PINE, VIRGINIA SCRUB PINE.

Tsuga canadensis (L.) Carrière   EASTERN HEMLOCK, CANADA HEMLOCK.

■ CUPRESSACEAE.   CYPRESS FAMILY

Juniperus communis L. **var. depressa** Pursh   GROUND JUNIPER.   (subsp. *depressa* (Pursh) Franco)
Juniperus virginiana L.   EASTERN RED-CEDAR.   (Incl. var. *crebra* Fernald & Griscom)
† **Taxodium distichum** (L.) Rich.   BALD-CYPRESS.   Often placed in a segregate family, the *Taxodiaceae*.
Thuja occidentalis L.   ARBOR VITAE, NORTHERN WHITE-CEDAR.

Order **Taxales**

■ TAXACEAE.   YEW FAMILY

Taxus canadensis Marshall   CANADA YEW.
† **Taxus cuspidata** Siebold & Zucc.   JAPANESE YEW.

Angiosperms or Flowering Plants: Dicotyledons (Dicots)

Tom S. Cooperrider, John J. Furlow,
and Allison W. Cusick

Phylum **MAGNOLIOPHYTA.**
ANGIOSPERMS

Class **Magnoliopsida.** DICOTS

Subclass **MAGNOLIIDAE**

Order **Magnoliales**

■ MAGNOLIACEAE.   Magnolia Family

Liriodendron tulipifera L.   Tulip-tree, Tulip-poplar, Yellow-poplar.
Magnolia acuminata (L.) L.   Cucumber-tree, Cucumber Magnolia.
†**Magnolia fraseri** Walter   Mountain Magnolia.
Magnolia macrophylla Michx.   Bigleaf Magnolia.
†**Magnolia stellata** (Siebold & Zucc.) Maxim.   Star Magnolia.
Magnolia tripetala (L.) L.   Umbrella Magnolia.
†**Magnolia × soulangiana** Soul.-Bod. (**Magnolia denudata** Desr. (M. *heptapeta* (Buc'hoz) Dandy) × **M. liliiflora** Desr. (M. *quinquepeta* (Buc'hoz) Dandy))   Saucer Magnolia, Chinese Magnolia.

■ ANNONACEAE.   Custard-apple Family

Asimina triloba (L.) Dunal   Pawpaw.

Order **Laurales**

■ CALYCANTHACEAE.   Strawberry-shrub Family

Calycanthus floridus L. **var. glaucus** (Willd.) Torr. & A. Gray   Sweet-shrub, Carolina-allspice.   (C. *fertilis* Walter)

■ LAURACEAE.   Laurel Family

Lindera benzoin (L.) Blume
   var. benzoin   Spicebush.
   var. pubescens (E. J. Palmer & Steyerm.) Rehder   Downy Spicebush.
Sassafras albidum (Nutt.) Nees   Sassafras. (Incl. var. *molle* (Raf.) Fernald)

Order **Piperales**

■ SAURURACEAE.   Lizard's-tail Family

Saururus cernuus L.   Lizard's-tail.

Order **Aristolochiales**

■ ARISTOLOCHIACEAE.   Birthwort Family

†**Aristolochia clematitis** L.   Birthwort.
Aristolochia serpentaria L.   Virginia-snakeroot.
†**Aristolochia tomentosa** Sims   Pipe-vine.
Asarum canadense L.   Wild Ginger.   (Incl. var. *reflexum* (E. P. Bicknell) B. L. Rob.)

Order **Nymphaeales**

■ NELUMBONACEAE.   Lotus-lily Family

Nelumbo lutea Willd.   American Lotus, American Water Lotus, Yellow Lotus.

†**Nelumbo nucifera** Gaertn.   SACRED LOTUS, INDIAN LOTUS.

■ NYMPHAEACEAE.   WATER-LILY FAMILY

Nuphar advena (Aiton) W. T. Aiton   SPATTER-DOCK, YELLOW POND-LILY.   (*N. lutea* Sm. subsp. *advena* (Aiton) Kartesz & Gandhi)
Nuphar variegata Durand   BULLHEAD-LILY.   (*N. lutea* Sm. subsp. *variegata* (Durand) E. O. Beal)
Nymphaea odorata Aiton   WHITE WATER-LILY, FRAGRANT WATER-LILY.   (Incl. subsp. *tuberosa* (Paine) Wiersema & Hellq., *N. tuberosa* Paine)

■ CABOMBACEAE.   WATER-SHIELD FAMILY

Brasenia schreberi J. F. Gmel.   WATER-SHIELD.
*****Cabomba caroliniana** A. Gray   FANWORT, WASHINGTON-GRASS.

■ CERATOPHYLLACEAE.   HORNWORT FAMILY

Ceratophyllum demersum L.   COONTAIL.
Ceratophyllum echinatum A. Gray   PRICKLY HORNWORT.

Order **Ranunculales**

■ RANUNCULACEAE.   BUTTERCUP FAMILY

Aconitum noveboracense A. Gray   NORTHERN MONKSHOOD, NEW YORK MONKSHOOD. (Treated by some authors as a part of *A. columbianum* Nutt.)
Aconitum uncinatum L.   SOUTHERN MONKS-HOOD.
Actaea pachypoda Elliott   WHITE BANEBERRY, DOLL'S-EYES.   (*A. alba* (L.) Mill.)
Actaea rubra (Aiton) Willd.   RED BANEBERRY, SNAKEBERRY.
†**Anemone blanda** Schott & Kotschy   GRECIAN WINDFLOWER.
Anemone canadensis L.   CANADA ANEMONE.
Anemone cylindrica A. Gray   PRAIRIE THIMBLEWEED, LONG-HEADED ANEMONE.

Anemone quinquefolia L.   WOOD ANEMONE.
Anemone virginiana L.   WOODLAND THIMBLEWEED, TALL ANEMONE.
Aquilegia canadensis L.   WILD COLUMBINE.
*****Aquilegia vulgaris** L.   GARDEN COLUMBINE, EUROPEAN COLUMBINE.
Caltha palustris L.   MARSH-MARIGOLD, COWSLIP.
Cimicifuga racemosa (L.) Nutt.   BLACK SNAKE-ROOT, BLACK COHOSH.   (*Actaea racemosa* L.)
Clematis occidentalis (Hornem.) DC.   PURPLE VIRGIN'S-BOWER, PURPLE CLEMATIS.
*****Clematis terniflora** DC.   YAM-LEAVED CLEMATIS.   (Incl. *C. dioscoreifolia* H. Lév. & Vaniot and *C. maximowicziana* Franch. & Sav.)
Clematis viorna L.   LEATHER-FLOWER.
Clematis virginiana L.   VIRGIN'S-BOWER, DEVIL'S-DARNING-NEEDLE.
†**Consolida ajacis** (L.) Schur   ROCKET LARKSPUR.   (*Delphinium ajacis* L.; treated by some authors as *C. ambigua* (L.) P. W. Ball & Heywood or as *Delphinium ambiguum* L.)
†**Consolida regalis** Gray   FORKING LARKSPUR.   (*Delphinium consolida* L.)
Coptis trifolia (L.) Salisb.   GOLDTHREAD.   (Incl. *C. groenlandica* (Oeder) Fernald)
Delphinium exaltatum Aiton   TALL LARKSPUR.
Delphinium tricorne Michx.   DWARF LARKSPUR.
†**Eranthis hyemalis** (L.) Salisb.   WINTER-ACONITE.
†**Helleborus viridis** L.   GREEN HELLEBORE.
Hepatica acutiloba DC.   SHARP-LOBED HEPATICA, SHARP-LOBED LIVERLEAF, SHARP-LOBED LIVERWORT.   (*H. nobilis* Mill. var. *acuta* (Pursh) Steyerm.; *Anemone acutiloba* (DC.) G. Lawson)
Hepatica americana (DC.) Ker Gawl.   ROUND-LOBED HEPATICA, ROUND-LOBED LIVERLEAF, ROUND-LOBED LIVERWORT.   (*H. nobilis* Mill. var. *obtusa* (Pursh) Steyerm.; *Anemone americana* (DC.) H. Hara)
Hepatica acutiloba × **H. americana**
Hydrastis canadensis L.   GOLDENSEAL.
Isopyrum biternatum (Raf.) Torr. & A. Gray   FALSE RUE-ANEMONE.   (*Enemion biternatum* Raf.)

Myosurus minimus L.   Mousetail.
†**Nigella damascena** L.   Love-in-a-mist.
Ranunculus abortivus L.   Small-flowered Crowfoot, Kidney-leaved Buttercup.
***Ranunculus acris** L.   Tall Buttercup, Common Buttercup.
Ranunculus alleghaniensis Britton   Allegheny Crowfoot.
Ranunculus ambigens S. Watson   Water-plantain Spearwort.
Ranunculus aquatilis L. **var. diffusus** With.   White Water-crowfoot, White Water-buttercup.   (*R. longirostris* Godr.)
†**Ranunculus arvensis** L.   Corn Crowfoot.
***Ranunculus bulbosus** L.   Bulbous Buttercup.
Ranunculus fascicularis Muhl. ex Bigelow   Early Buttercup, Thick-rooted Buttercup, Prairie Buttercup.
***Ranunculus ficaria** L.   Lesser Celandine, Pilewort, Golden-stars.
Ranunculus flabellaris Raf.   Yellow Water-buttercup, Yellow Water-crowfoot.
Ranunculus hispidus Michx.
  var. caricetorum (Greene) T. Duncan   Northern Swamp Buttercup.   (*R. septentrionalis* Poir. var. *caricetorum* (Greene) Fernald)
  var. hispidus   Hispid Buttercup.
  var. nitidus (Chapm.) T. Duncan   Swamp Buttercup.   (Incl. *R. carolinianus* DC. and *R. septentrionalis* Poir. var. *septentrionalis*)
Ranunculus micranthus Nutt.   Small-flowered Crowfoot.
***Ranunculus parviflorus** L.   Stickseed Crowfoot, Small-flowered Buttercup.
Ranunculus pensylvanicus L. f.   Bristly Buttercup, Bristly Crowfoot.
Ranunculus pusillus Poir.   Low Spearwort, Dwarf Crowfoot.
Ranunculus recurvatus Poir.   Hooked Crowfoot.
***Ranunculus repens** L.   Creeping Buttercup.   (Incl. var. *pleniflorus* Fernald)
Ranunculus sceleratus L.   Cursed Crowfoot.
***Ranunculus testiculatus** Crantz   Bur Buttercup.   (*Ceratocephalus testiculatus* (Crantz) Roth)
Thalictrum dasycarpum Fisch. & Avé-Lall.   Purple Meadow-rue.
Thalictrum dioicum L.   Early Meadow-rue.
Thalictrum pubescens Pursh   Tall Meadow-rue.   (Incl. *T. polygamum* Muhl. ex Spreng.)
Thalictrum revolutum DC.   Skunk Meadow-rue, Waxy Meadow-rue.
Thalictrum thalictroides (L.) A. J. Eames & B. Boivin   Rue-anemone.   (*Anemonella thalictroides* (L.) Spach)
Trollius laxus Salisb.   Spreading Globeflower.
†**Xanthorhiza simplicissima** Marshall   Yellow-root.

■ BERBERIDACEAE.   Barberry Family

***Berberis thunbergii** DC.   Japanese Barberry.
***Berberis vulgaris** L.   Common Barberry.
***Berberis × ottawensis** C. K. Schneid. (**Berberis thunbergii × B. vulgaris**)   Ottawa Barberry.
Caulophyllum thalictroides (L.) Michx.
  var. giganteum Farw.   Giant Blue Cohosh, Large-flowered Blue Cohosh.   (*C. giganteum* (Farw.) Loconte and W. H. Blackw.)
  var. thalictroides   Blue Cohosh, Papoose-root.
Jeffersonia diphylla (L.) Pers.   Twinleaf.
†**Mahonia aquifolium** (Pursh) Nutt.   Oregon-grape.   (*Berberis aquifolium* Pursh)
Podophyllum peltatum L.   Mayapple, Mandrake.

■ LARDIZABALACEAE.   Lardizabala Family

†**Akebia quinata** (Houtt.) Decne.   Five-leaf Akebia, Chocolate-vine.

■ MENISPERMACEAE.   Moonseed Family

Menispermum canadense L.   Moonseed.

Order **Papaverales**

- PAPAVERACEAE. Poppy Family

†**Argemone albiflora** Hornem. White Prickly-
poppy.
†**Argemone mexicana** L. Mexican Prickly-
poppy.
*__Chelidonium majus__ L. Celandine.
†**Eschscholzia californica** Cham. California-
poppy.
†**Macleaya cordata** (Willd.) R. Br. Plume-
poppy.
†**Papaver argemone** L. Red Poppy.
*__Papaver dubium__ L. Corn Poppy.
†**Papaver rhoeas** L. Field Poppy.
†**Papaver somniferum** L. Opium Poppy.
Sanguinaria canadensis L. Bloodroot.
(Incl. var. *rotundifolia* (Greene) Fedde)
Stylophorum diphyllum (Michx.) Nutt.
Celandine-poppy, Wood-poppy.

- FUMARIACEAE. Fumitory Family

Adlumia fungosa (Aiton) Greene ex Britton,
Sterns & Poggenb. Allegheny-vine,
Mountain-fringe, Climbing Fumitory.
†**Corydalis aurea** Willd. Golden Corydalis.
Corydalis flavula (Raf.) DC. Yellow-
harlequin, Yellow Corydalis, Short-
spurred Corydalis.
Corydalis sempervirens (L.) Pers. Rock-
harlequin, Pink Corydalis.
Dicentra canadensis (Goldie) Walp. Squirrel-
corn.
Dicentra cucullaria (L.) Bernh. Dutchman's-
breeches.
†**Dicentra eximia** (Ker Gawl.) Torr. Turkey-
corn, Eastern Bleeding-heart, Wild
Bleeding-heart.
†**Fumaria officinalis** L. Common Fumitory,
Earth-smoke.

Subclass **HAMAMELIDAE**

Order **Hamamelidales**

- CERCIDIPHYLLACEAE. Cercidiphyllum
Family

†**Cercidiphyllum japonicum** Siebold & Zucc. ex
J. J. Hoffm. & H. Schult. Katsura-tree.

- PLATANACEAE. Plane-tree Family

Platanus occidentalis L. Sycamore, American
Plane-tree.

- HAMAMELIDACEAE. Witch-hazel
Family

Hamamelis virginiana L. Witch-hazel.
(Incl. var. *parviflora* Nutt.)
Liquidambar styraciflua L. Sweet Gum.

Order **Urticales**

- ULMACEAE. Elm Family

Celtis occidentalis L. Hackberry, Northern
Hackberry. (Incl. var. *pumila* (Pursh)
A. Gray)
Celtis tenuifolia Nutt. Dwarf Hackberry.
(Incl. var. *georgiana* (Small) Fernald & B. G.
Schub.)
Ulmus americana L. American Elm, White
Elm.
†**Ulmus minor** Mill. English Elm. (Incl.
U. procera Salisb.)
†**Ulmus pumila** L. Siberian Elm, Chinese
Elm.
Ulmus rubra Muhl. Slippery Elm, Red Elm.
Ulmus thomasii Sarg. Rock Elm, Cork Elm.
†**Zelkova serrata** (Thunb.) Makino Japanese
Zelkova.

- CANNABACEAE. Hemp Family

†**Cannabis sativa** L. Hemp.
*__Humulus japonicus__ Siebold & Zucc. Japanese
Hops—or Hop.
Humulus lupulus L. Common Hops—or Hop.
var. lupuloides E. Small
*__var. lupulus__
var. pubescens E. Small

- MORACEAE. Mulberry Family

†**Broussonetia papyrifera** (L.) L'Hér. ex Vent.
Paper-mulberry.

*Fatoua villosa** (Thunb.) Nakai MULBERRY-WEED, HAIRY CRABWEED.
*Maclura pomifera** (Raf.) C. K. Schneid. OSAGE-ORANGE.
*Morus alba** L. WHITE MULBERRY.
Morus rubra L. RED MULBERRY.

■ URTICACEAE. NETTLE FAMILY

Boehmeria cylindrica (L.) Sw. FALSE NETTLE, BOG-HEMP.
Laportea canadensis (L.) Wedd. WOOD-NETTLE.
Parietaria pensylvanica Muhl. ex Willd. PELLITORY.
Pilea fontana (Lunnell) Rydb. MARSH CLEARWEED.
Pilea pumila (L.) A. Gray COMMON CLEARWEED, RICHWEED.
Urtica chamaedryoides Pursh SPRING NETTLE.
Urtica dioica L. STINGING NETTLE.
　*var. dioica** EUROPEAN STINGING NETTLE.
　var. procera (Muhl. ex Willd.) Wedd. AMERICAN STINGING NETTLE. (Incl. var. *gracilis* (Aiton) C. L. Hitchc., subsp. *gracilis* (Aiton) Selander)

Order **Juglandales**

■ JUGLANDACEAE. WALNUT FAMILY

Carya cordiformis (Wangenh.) K. Koch BITTERNUT HICKORY.
Carya glabra (Mill.) Sweet PIGNUT HICKORY. (Incl. var. *megacarpa* Sarg.)
Carya illinoinensis (Wangenh.) K. Koch PECAN. (Treated by some authors as *C. oliviformis* (Michx.) Nutt.)
Carya laciniosa (F. Michx.) Loudon SHELLBARK HICKORY, KINGNUT.
Carya ovalis (Wangenh.) Sarg. SWEET PIGNUT HICKORY. (Incl. var. *mollis* (Ashe) Sudw. and var. *obcordata* (Muhl. & Willd.) Sarg.; treated by some authors as a part of *C. glabra*)
Carya ovata (Mill.) K. Koch SHAGBARK HICKORY.
Carya tomentosa (Poir.) Nutt. MOCKERNUT HICKORY. (Treated by some authors as *C. alba* (L.) K. Koch)
Juglans cinerea L. BUTTERNUT, WHITE WALNUT.

Juglans nigra L. BLACK WALNUT.

Order **Myricales**

■ MYRICACEAE. BAYBERRY FAMILY

Comptonia peregrina (L.) J. M. Coult. SWEET-FERN. (Incl. var. *aspleniifolia* (L.) Fernald)
Myrica pensylvanica Loisel. BAYBERRY, NORTHERN BAYBERRY.

Order **Fagales**

■ FAGACEAE. BEECH FAMILY

Castanea dentata (Marsh.) Borkh. AMERICAN CHESTNUT.
†**Castanea pumila** (L.) Mill. CHINQUAPIN.
Fagus grandifolia Ehrh. AMERICAN BEECH.
　var. caroliniana** (Loudon) Fernald & Rehder WHITE BEECH.
　var. grandifolia RED BEECH.
†**Fagus sylvatica** L. EUROPEAN BEECH.
Quercus alba L. WHITE OAK.
Quercus bicolor Willd. SWAMP WHITE OAK.
Quercus coccinea Münchh. SCARLET OAK. (Incl. var. *tuberculata* Sarg.)
Quercus ellipsoidalis E. J. Hill NORTHERN PIN OAK, JACK OAK. (Treated by some authors as a part of *Q. coccinea*)
Quercus falcata Michx. SOUTHERN RED OAK, SPANISH OAK.
Quercus imbricaria Michx. SHINGLE OAK.
Quercus macrocarpa Michx. BUR OAK, MOSSY-CUP OAK.
Quercus marilandica Münchh. BLACKJACK OAK.
Quercus muehlenbergii Engelm. CHINQUAPIN OAK, YELLOW CHESTNUT OAK.
Quercus palustris Münchh. PIN OAK.
Quercus prinus L. ROCK CHESTNUT OAK. (Incl. *Q. montana* Willd.)
†**Quercus robur** L. ENGLISH OAK.
Quercus rubra L. RED OAK, NORTHERN RED OAK. (Incl. var. *ambigua* (A. Gray) Fernald, var. *borealis* (F. Michx.) Farw., and *Q. borealis* F. Michx. var. *borealis* and var. *maxima* (Marshall) Ashe, *Q. maxima* (Marshall) Ashe)

Quercus shumardii Buckley SHUMARD OAK, SHUMARD RED OAK.   (Incl. var. *schneckii* (Britton) Sarg., *Q. schneckii* Britton)

Quercus stellata Wangenh.   POST OAK.

Quercus velutina Lam.   BLACK OAK.

Quercus × bebbiana C. K. Schneid. (**Quercus alba × Q. macrocarpa**) BEBB'S OAK.

Quercus × bushii Sarg. (**Quercus marilandica × Q. velutina**) BUSH'S OAK.

Quercus × exacta Trel. (**Quercus imbricaria × Q. palustris**) TRELEASE'S OAK.

Quercus × fontana Laughlin (**Quercus coccinea × Q. velutina**) FONTANA'S OAK.

Quercus × hawkinsiae Sudw. (**Quercus rubra × Q. velutina**) HAWKINS' OAK.

Quercus × jackiana C. K. Schneid. (**Quercus alba × Q. bicolor**) SCHNEIDER'S OAK.

Quercus × leana Nutt. (**Quercus imbricaria × Q. velutina**) LEA'S OAK.

Quercus × mutabilis E. J. Palmer & Steyerm. (**Quercus palustris × Q. shumardii**) VARIABLE OAK.

Quercus × runcinata (A. DC.) Engelm. (**Quercus imbricaria × Q. rubra**) SAWTOOTH OAK.

Quercus × saulii C. K. Schneid. (**Quercus alba × Q. prinus**) SAUL'S OAK.

Quercus × schuettei Trel. (**Quercus bicolor × Q. macrocarpa**) SCHUETTE'S OAK.   (Incl. *Q. × hillii* Trel.)

Quercus × tridentata (A. DC.) Trel. (**Quercus imbricaria × Q. marilandica**) THREE-TOOTHED OAK.

Quercus alba × Q. muehlenbergii

Quercus coccinea × Q. imbricaria

■ BETULACEAE. BIRCH FAMILY

*Alnus glutinosa (L.) Gaertn.   BLACK ALDER.

Alnus incana (L.) Moench **subsp. rugosa** (Du Roi) R. T. Clausen   SPECKLED ALDER, TAG ALDER.   (*A. rugosa* (Du Roi) Spreng.; incl. var. *americana* Regel)

Alnus serrulata (Aiton) Willd.   SMOOTH ALDER, COMMON ALDER.

Alnus incana × A. serrulata

Betula alleghaniensis Britton   YELLOW BIRCH.   (Incl. var. *macrolepis* (Fernald) Brayshaw; treated by some authors as *B. lutea* F. Michx.)

Betula lenta L.   SWEET BIRCH, CHERRY BIRCH, BLACK BIRCH.

Betula nigra L.   RIVER BIRCH, RED BIRCH.

†Betula papyrifera Marshall   PAPER BIRCH, CANOE BIRCH.

*Betula pendula Roth   EUROPEAN WEEPING BIRCH, EUROPEAN WHITE BIRCH.   (Treated by some authors as *B. alba* L.)

Betula populifolia Marshall   GRAY BIRCH, WIRE BIRCH.

†Betula pubescens Ehrh.   DOWNY BIRCH.   (Treated by some authors as *B. alba* L.)

Betula pumila L.   BOG BIRCH, SWAMP BIRCH, DWARF BIRCH.

Betula × purpusii C. K. Schneid. (**Betula alleghaniensis × B. pumila**) PURPUS' BIRCH.

Carpinus caroliniana Walter **subsp. virginiana** (Marshall) Furlow   BLUE-BEECH, AMERICAN HORNBEAM, MUSCLEWOOD, IRONWOOD.

Corylus americana Walter   AMERICAN HAZEL, AMERICAN HAZELNUT.

Corylus cornuta Marshall   BEAKED HAZEL, BEAKED HAZELNUT.

Ostrya virginiana (Mill.) K. Koch   HOP-HORNBEAM, IRONWOOD.

Subclass **CARYOPHYLLIDAE**

Order **Caryophyllales**

■ PHYTOLACCACEAE. POKEWEED FAMILY

Phytolacca americana L.   POKEWEED, POKEBERRY, POKE.

■ NYCTAGINACEAE. FOUR-O'CLOCK FAMILY

†Mirabilis albida (Walter) Heimerl   PALE UMBRELLA-WORT.

†Mirabilis hirsuta (Pursh) MacMill.   HAIRY UMBRELLA-WORT.

†Mirabilis jalapa L.   FOUR-O'CLOCK, MARVEL-OF-PERU.

*Mirabilis nyctaginea (Michx.) MacMill.   HEART-LEAVED UMBRELLA-WORT, WILD FOUR-O'CLOCK.

■ AIZOACEAE. Fig-marigold Family

† **Tetragonia tetragonioides** (Pall.) Kuntze New Zealand-spinach.

■ CACTACEAE. Cactus Family

Opuntia humifusa (Raf.) Raf. Common Prickly-pear, Eastern Prickly-pear.

****Opuntia macrorhiza** Engelm. Plains Prickly-pear. (Incl. *O. tortispina* Engelm. & Bigel.)

■ CHENOPODIACEAE. Goosefoot Family

****Atriplex hortensis** L. Garden Orache—also spelled Orach.

****Atriplex patula** L. Spear-scale, Halberd-leaved Orache. (Incl. *A. subspicata* (Nutt.) Rydb.)

****Atriplex prostrata** Boucher ex DC. Prostrate Orache. (Treated by some authors as *A. patula* var. *hastata* (L.) A. Gray)

† **Atriplex rosea** L. Red Orache.

****Chenopodium album** L. Lamb's-quarters, White Goosefoot. (Incl. var. *lanceolatum* (Muhl. ex Willd.) Coss. & Germ.)

****Chenopodium ambrosioides** L. Mexican-tea, Wormseed.

Chenopodium berlandieri Moq. Pitseed Goosefoot.
 var. bushianum (Aellen) Cronquist
 var. zschackei (Murr.) Murr. ex Asch. (subsp. *zschackei* (Murr.) A. Zobel)

****Chenopodium botrys** L. Jerusalem-oak, Feather-geranium.

Chenopodium capitatum (L.) Asch. Strawberry-blite.

****Chenopodium glaucum** L. Oak-leaved Goosefoot.

Chenopodium hybridum L. **var. gigantosper-mum** (Aellen) Rouleau Maple-leaved Goosefoot. (subsp. *gigantospermum* (Aellen) Hultén)

† **Chenopodium incanum** (S. Watson) A. Heller Hoary Goosefoot.

Chenopodium leptophyllum (Moq.) Nutt. ex S. Watson Slender Goosefoot, Narrow-leaved Goosefoot. (Incl. *C. subglabrum* (S. Watson) A. Nelson)

Chenopodium missouriense Aellen Missouri Goosefoot.

****Chenopodium murale** L. Nettle-leaved Goosefoot.

† **Chenopodium polyspermum** L. Many-seeded Goosefoot.

Chenopodium pratericola Rydb. Field Goosefoot.

† **Chenopodium pumilio** R. Br. Dwarf Goosefoot.

****Chenopodium rubrum** L. Alkali-blite. (Incl. var. *humile* (Hook.) S. Watson, *C. humile* Hook.)

Chenopodium standleyanum Aellen Woodland Goosefoot.

****Chenopodium urbicum** L. City Goosefoot.

† **Chenopodium vulvaria** L. Fetid Goosefoot, Stinking Goosefoot.

† **Corispermum americanum** (Nutt.) Nutt. American Bugseed. (*C. hyssopifolium* L. var. *americanum* Nutt.)

****Corispermum pallasii** Steven Narrow-winged Bugseed. (Treated by some authors as *C. nitidum* Kit. ex Schult.)

****Cycloloma atriplicifolium** (Spreng.) J. M. Coult. Winged Pigweed.

† **Kochia scoparia** (L.) Roth ex Schrad. Summer-cypress. (Incl. var. *culta* Farw.)

† **Monolepis nuttalliana** (Schult.) Greene Poverty-weed.

****Salicornia europaea** L. Samphire, Chicken-claws.

****Salsola collina** Pall. Slender Russian-thistle.

† **Salsola tragus** L. Russian-thistle. (Treated by some authors as *S. kali* L. var. *tenuifolia* Tausch)

† **Spinacia oleracea** L. Spinach.

****Suaeda calceoliformis** (Hook.) Moq. Plains Sea-blite.

■ AMARANTHACEAE. Amaranth Family

† **Achyranthes japonica** (Miq.) Nakai Japanese Chaff-flower.

Amaranthus albus L. Tumbleweed.

*Amaranthus blitoides S. Watson Mat
Amaranth. (Treated by some authors as
A. *graecizans* L.)

† Amaranthus blitum L. Weedy Amaranth.
(Incl. A. *lividus* L.)

*Amaranthus cruentus L. Red Amaranth,
Purple Amaranth, Prince's-feather.

*Amaranthus hybridus L. Smooth Pigweed,
Green Amaranth.

*Amaranthus palmeri S. Watson Careless-
weed.

† Amaranthus powellii S. Watson Powell's
Amaranth.

*Amaranthus retroflexus L. Redroot, Rough
Pigweed.

Amaranthus rudis J. D. Sauer Western
Pigweed. (Treated by some authors as
A. *tamariscinus* Nutt.)

*Amaranthus spinosus L. Spiny Amaranth,
Thorny Amaranth.

Amaranthus tuberculatus (Moq.) J. D. Sauer
Tubercled Amaranth, Tubercled Water-
hemp.

† Celosia argentea L. var. cristata (L.) Kuntze
Cockscomb.

Froelichia floridana (Nutt.) Moq. var. campestris
(Small) Fernald Common Cottonweed.

Froelichia gracilis (Hook.) Moq. Slender
Cottonweed.

† Gomphrena globosa L. Globe Amaranth.

■ PORTULACACEAE. Purslane Family

Claytonia caroliniana Michx. Carolina
Spring-beauty.

Claytonia virginica L. Spring-beauty.

† Portulaca grandiflora Hook. Moss-rose.

*Portulaca oleracea L. Common Purslane.

■ MOLLUGINACEAE. Carpet-weed Family

*Mollugo verticillata L. Carpet-weed.

■ CARYOPHYLLACEAE. Pink Family

*Agrostemma githago L. Corn-cockle.

*Arenaria serpyllifolia L. Thyme-leaved
Sandwort.

Cerastium arvense L. Field Chickweed.

*Cerastium brachypetalum Pers. Gray
Chickweed.

† Cerastium dubium (T. Bastard) Guépin
Three-styled Chickweed, Doubtful
Chickweed.

*Cerastium fontanum Baumg. Common
Mouse-ear Chickweed. (Treated by some
authors as a part of C. *vulgatum* L.)

*Cerastium glomeratum Thuill. Clammy
Chickweed. (Treated by some authors as
a part of C. *viscosum* L.)

Cerastium nutans Raf. Nodding Mouse-ear
Chickweed. (Incl. C. *longepedunculatum*
Muhl. ex Britton)

*Cerastium pumilum Curtis Curtis' Mouse-
ear Chickweed.

*Cerastium semidecandrum L. Small Mouse-
ear Chickweed.

† Cerastium tomentosum L. Snow-in-summer.

*Dianthus armeria L. Deptford Pink.

† Dianthus barbatus L. Sweet William.

† Dianthus caryophyllus L. Carnation.

† Dianthus deltoides L. Maiden Pink.

† Dianthus plumarius L. Garden Pink, Grass
Pink.

† Gypsophila muralis L. Cushion Baby's-
breath.

† Gypsophila paniculata L. Baby's-breath.

† Gypsophila scorzonerifolia Ser. Garden
Baby's-breath.

*Holosteum umbellatum L. Jagged
Chickweed.

*Lychnis coronaria (L.) Desr. Mullein-pink,
Rose Campion.

† Lychnis flos-cuculi L. Ragged-robin,
Cuckoo-flower.

† Lychnis viscaria L. German Catchfly.

Minuartia michauxii (Fenzl) Farw. var. michauxii
Rock Sandwort. (*Arenaria stricta* Michx.)

Minuartia patula (Michx.) Mattf. Spreading
Sandwort, Slender Sandwort. (*Arenaria
patula* Michx.)

Moehringia lateriflora (L.) Fenzl Grove
Sandwort. (*Arenaria lateriflora* L.)

† Moehringia trinervia (L.) Clairv. Three-
nerved Sandwort.

*Myosoton aquaticum** (L.) Moench WATER CHICKWEED, GIANT CHICKWEED. (*Stellaria aquatica* (L.) Scop.)

Paronychia canadensis (L.) A. W. Wood COMMON FORKED-CHICKWEED.

Paronychia fastigiata (Raf.) Fernald STIFF FORKED-CHICKWEED.

†**Petrorhagia prolifera** (L.) P. W. Ball & Heywood CHILDING-PINK. (*Dianthus prolifer* L., *Tunica prolifera* (L.) Scop.)

†**Petrorhagia saxifraga** (L.) Link SAXIFRAGE-PINK. (*Tunica saxifraga* (L.) Scop.)

Sagina decumbens (Elliott) Torr. & A. Gray WESTERN PEARLWORT, SOUTHERN ·PEARLWORT.

†**Sagina japonica** (Sw.) Ohwi JAPANESE PEARLWORT.

*Sagina procumbens** L. ARCTIC PEARLWORT, BIRD'S-EYE PEARLWORT.

*Saponaria officinalis** L. SOAPWORT, BOUNCING BET.

*Scleranthus annuus** L. ANNUAL KNAWEL.

Silene antirrhina L. SLEEPY CATCHFLY.

†**Silene armeria** L. GARDEN CATCHFLY, SWEET-WILLIAM CATCHFLY, NONE-SO-PRETTY.

Silene caroliniana Walter
  var. pensylvanica (Michx.) Fernald CAROLINA CATCHFLY. (subsp. *pensyl-vanica* (Michx.) R. T. Clausen)
  var. wherryi (Small) Fernald WHERRY'S CATCHFLY. (subsp. *wherryi* (Small) R. T. Clausen)

†**Silene conica** L. STRIATE CATCHFLY.

*Silene csereii** Baumg. BALKAN CAMPION.

*Silene dichotoma** Ehrh. FORKED CATCHFLY.

†**Silene dioica** (L.) Clairv. RED CAMPION, RED COCKLE. (*Lychnis dioica* L.)

*Silene latifolia** Poir. WHITE CAMPION, WHITE COCKLE. (Incl. *S. pratensis* (Raf.) Godr. & Gren.; *Lychnis alba* Mill.)

Silene nivea (Nutt.) Muhl. ex Otth SNOWY CAMPION.

*Silene noctiflora** L. NIGHT-BLOOMING CATCHFLY, STICKY-COCKLE.

Silene regia Sims ROYAL CATCHFLY.

Silene rotundifolia Nutt. ROUND-LEAVED CATCHFLY.

Silene stellata (L.) W. T. Aiton STARRY CAMPION, WIDOW'S-FRILL.

Silene virginica L. FIRE-PINK.

*Silene vulgaris** (Moench) Garcke BLADDER CAMPION. (Incl. *S. cucubalis* Wibel and *S. latifolia* (Mill.) Britten & Rendle—not *S. latifolia* Poir., listed above)

*Spergula arvensis** L. CORN SPURRY—also spelled SPURREY.

*Spergularia marina** (L.) Griseb. SALT-MARSH SAND SPURRY.

*Spergularia media** (L.) C. Presl ex Griseb. SALT SAND SPURRY.

*Spergularia rubra** (L.) J. Presl & C. Presl ROADSIDE SAND SPURRY, RED SAND SPURRY.

†**Stellaria alsine** Grimm BOG STITCHWORT.

Stellaria corei Shinners KENTUCKY CHICKWEED, CORE'S CHICKWEED. (*S. pubera* var. *sylvatica* (Bég.) Weath.)

*Stellaria graminea** L. COMMON STITCHWORT.

†**Stellaria holostea** L. GREATER STITCHWORT, EASTER-BELL.

Stellaria longifolia Muhl. ex Willd. LONG-LEAVED STITCHWORT.

*Stellaria media** (L.) Vill. COMMON CHICKWEED.

*Stellaria pallida** (Dumort.) Piré LESSER CHICKWEED.

*Stellaria palustris** (Murray) Retz. MEADOW STARWORT.

Stellaria pubera Michx. STAR CHICKWEED, GREATER CHICKWEED.

†**Vaccaria hispanica** (Mill.) Rauschert COW-HERB. (*Saponaria vaccaria* L.)

Order **Polygonales**

- POLYGONACEAE. BUCKWHEAT FAMILY or SMARTWEED FAMILY

†**Fagopyrum esculentum** Moench BUCKWHEAT. (*F. sagittatum* Gilib.)

*Polygonum achoreum** S. F. Blake STRIATE KNOTWEED.

Polygonum amphibium L. WATER SMARTWEED.
  var. emersum Michx.
  var. stipulaceum N. Coleman

*Polygonum arenastrum Jord. ex Boreau
DOORYARD KNOTWEED, OVAL-LEAVED
KNOTWEED. (*P. aviculare* var. *arensatrum*
(Jord. ex Boreau) Rouy)

Polygonum arifolium L. **var. pubescens** (Keller)
Fernald HALBERD-LEAVED TEAR-THUMB.

*Polygonum aviculare L. COMMON KNOTWEED,
PROSTRATE KNOTWEED.

Polygonum careyi Olney CAREY'S
SMARTWEED.

*Polygonum cespitosum Blume **var. longisetum**
(Bruyn) Steward LONG-BRISTLED
SMARTWEED, BRISTLY LADY'S-THUMB,
TUFTED KNOTWEED.

Polygonum cilinode Michx. FRINGED FALSE
BUCKWHEAT, FRINGED BINDWEED, MOUNTAIN
BINDWEED, BLACK-FRINGED KNOTWEED.

*Polygonum convolvulus L. FALSE
BUCKWHEAT, BLACK BINDWEED, WILD
BUCKWHEAT.

*Polygonum cuspidatum Siebold & Zucc.
JAPANESE KNOTWEED, MEXICAN-BAMBOO.

Polygonum erectum L. ERECT KNOTWEED.

Polygonum hydropiper L. WATER-PEPPER,
MARSH-PEPPER SMARTWEED.

Polygonum hydropiperoides Michx.
var. hydropiperoides MILD WATER-PEPPER,
FALSE WATER-PEPPER.
var. setaceum (Baldwin) Gleason BRISTLY
SMARTWEED. (*P. setaceum* Baldwin)

Polygonum lapathifolium L. WILLOW-WEED,
NODDING SMARTWEED.

*Polygonum orientale L. PRINCE'S-FEATHER,
KISS-ME-OVER-THE-GARDEN-GATE.

Polygonum pensylvanicum L. PINKWEED,
PENNSYLVANIA SMARTWEED.
var. eglandulosum Myers
var. pensylvanicum

*Polygonum perfoliatum L. MILE-A-MINUTE.

*Polygonum persicaria L. LADY'S-THUMB,
HEART'S-EASE.

Polygonum punctatum Elliott WATER
SMARTWEED, DOTTED SMARTWEED.

Polygonum ramosissimum Michx. BUSHY
KNOTWEED.

Polygonum robustius (Small) Fernald COARSE
SMARTWEED, ROBUST SMARTWEED.
(*P. punctatum* var. *robustius* Small)

*Polygonum sachalinense F. W. Schmidt ex
Maxim. GIANT KNOTWEED, SACHALINE.

Polygonum sagittatum L. ARROW-LEAVED
TEAR-THUMB.

Polygonum scandens L. CLIMBING FALSE
BUCKWHEAT, HEDGE SMARTWEED.
var. cristatum (Engelm. & A. Gray) Gleason
(*P. cristatum* Engelm. & A. Gray)
*var. dumetorum (L.) Gleason
var. scandens

Polygonum tenue Michx. SLENDER
KNOTWEED.

Polygonum virginianum L. JUMPSEED, VIRGINIA
KNOTWEED.

†Rheum rhabarbarum L. RHUBARB. (Treated
by some authors as *R. rhaponticum* L.)

†Rumex acetosa L. EURASIAN GREEN SORREL.

*Rumex acetosella L. SHEEP SORREL, RED
SORREL.

Rumex altissimus A. W. Wood PALE DOCK.

†Rumex conglomeratus Murray CLUSTERED
DOCK.

*Rumex crispus L. CURLY DOCK, YELLOW
DOCK.

*Rumex maritimus L. **var. fueginus** (Phil.)
Dusén GOLDEN DOCK.

*Rumex obtusifolius L. BITTER DOCK.

Rumex orbiculatus A. Gray GREAT WATER
DOCK.

†Rumex patientia L. PATIENCE DOCK, MONK'S-
RHUBARB.

*Rumex salicifolius Weinm. WILLOW-LEAVED
DOCK. (Incl. *R. mexicanus* Meisn.)

Rumex verticillatus L. WATER DOCK, SWAMP
DOCK.

Subclass **DILLENIIDAE**

Order **Theales**

■ ELATINACEAE. WATERWORT FAMILY

Elatine triandra Schkuhr **var. brachysperma**
(A. Gray) Fassett WATERWORT, ELATINE.
(*E. brachysperma* A. Gray)

- CLUSIACEAE or GUTTIFERAE.
  Mangosteen Family or St. John's-wort
  Family

Hypericum boreale (Britton) E. P. Bicknell
  Northern St. John's-wort. (*H. mutilum*
  subsp. *boreale* (Britton) J. M. Gillett)

Hypericum canadense L. Canada St. John's-
  wort.

Hypericum denticulatum Walter **var. acuti-
  folium** (Elliott) S. F. Blake Coppery
  St. John's-wort. (*H. harperi* R. Keller)

Hypericum drummondii (Grev. & Hook.)
  Torr. & A. Gray Nits-and-lice.

Hypericum ellipticum Hook. Elliptic-leaved
  St. John's-wort, Few-flowered St. John's-
  wort.

Hypericum gentianoides (L.) Britton, Sterns &
  Poggenb. Orange-grass, Pineweed.

Hypericum gymnanthum Engelm. & A. Gray
  Clasping-leaved St. John's-wort, Least
  St. John's-wort.

Hypericum hypericoides (L.) Crantz **var.
  multicaule** (Michx. ex Willd.) Fosberg
  St. Andrew's-cross. (subsp. *multicaule*
  (Michx. ex Willd.) N. Robson; incl. *H. stragu-
  lum* W. P. Adams & N. Robson; *Ascyrum
  hypericoides* L. var. *multicaule* (Michx. ex
  Willd.) Fernald)

Hypericum kalmianum L. Kalm's St. John's-
  wort.

Hypericum majus (A. Gray) Britton Tall
  St. John's-wort.

Hypericum mutilum L. Small-flowered
  St. John's-wort.

*****Hypericum perforatum** L. Common
  St. John's-wort.

Hypericum prolificum L. Shrubby St. John's-
  wort. (Incl. *H. spathulatum* (Spach) Steud.)

Hypericum punctatum Lam. Spotted
  St. John's-wort.

Hypericum pyramidatum Aiton Great
  St. John's-wort. (Treated by some authors
  as a part of *H. ascyron* L.)

Hypericum sphaerocarpum Michx. Round-
  fruited St. John's-wort.

Hypericum canadense × **H. mutilum**

Triadenum tubulosum (Walter) Gleason
  (*Hypericum tubulosum* Walter)
  var. tubulosum Large Marsh St. John's-
  wort. (*H. tubulosum* var. *tubulosum*)
  var. walteri (J. G. Gmel.) Cooperr. Walter's
  St. John's-wort. (*T. walteri* (J. G. Gmel.)
  Gleason; *H. tubulosum* var. *walteri* (J. G.
  Gmel.) Lott)

Triadenum virginicum (L.) Raf. (*Hypericum vir-
  ginicum* L.)
  var. fraseri (Spach) Cooperr. Fraser's
  St. John's-wort. (*T. fraseri* (Spach)
  Gleason; *H. virginicum* var. *fraseri* (Spach)
  Fernald)
  var. virginicum Marsh St. John's-wort.
  (*H. virginicum* var. *virginicum*)

Order **Malvales**

- TILIACEAE. Linden Family

Tilia americana L. Basswood. (Incl. *T. ne-
  glecta* Spach)
†**Tilia europaea** L. European Linden.
  (Treated by some authors as *T.* × *vulgaris*
  Hayne)
Tilia heterophylla Vent. White Basswood.
  (*T. americana* var. *heterophylla* (Vent.) Loudon)

- MALVACEAE. Mallow Family

*****Abutilon theophrasti** Medik. Velvet-leaf.
†**Alcea rosea** L. Hollyhock. (*Althaea
  rosea* (L.) Cav.)
*****Althaea officinalis** L. Marsh-mallow.
†**Anoda cristata** (L.) Schltdl. Spurred
  Anoda.
Hibiscus laevis All. Halberd-leaved Rose-
  mallow, Soldier Rose-mallow. (Incl.
  H. militaris Cav.)
Hibiscus moscheutos L. Swamp Rose-
  mallow, Common Rose-mallow. (Incl.
  H. palustris L.)
†**Hibiscus syriacus** L. Rose-of-Sharon.
*****Hibiscus trionum** L. Flower-of-an-hour.
†**Malva alcea** L. Vervain Mallow.
*****Malva moschata** L. Musk Mallow.

***Malva neglecta** Wallr.   COMMON MALLOW, CHEESES.

***Malva pusilla** Sm.   ROUND-LEAVED MALLOW, DWARF MALLOW.   (Until recently, plants of this species were called M. *rotundifolia* L.)

†**Malva sylvestris** L.   HIGH MALLOW.

†**Malva verticillata** L.   WHORLED MALLOW.

Napaea dioica L.   GLADE-MALLOW.

Sida hermaphrodita (L.) Rusby   TALL SIDA, VIRGINIA-MALLOW.

***Sida spinosa** L.   PRICKLY SIDA.

Order **Nepenthales**

■ SARRACENIACEAE.   PITCHER-PLANT FAMILY

Sarracenia purpurea L.   PITCHER-PLANT, NORTHERN PITCHER-PLANT.

■ DROSERACEAE.   SUNDEW FAMILY

Drosera intermedia Hayne   NARROW-LEAVED SUNDEW, SPATHULATE-LEAVED SUNDEW, SPATULATE-LEAVED SUNDEW.

Drosera rotundifolia L.   ROUND-LEAVED SUNDEW.

Order **Violales**

■ CISTACEAE.   ROCK-ROSE FAMILY

Helianthemum bicknellii Fernald   PLAINS FROSTWEED.

Helianthemum canadense (L.) Michx.   CANADA FROSTWEED.

Hudsonia tomentosa Nutt.   BEACH-HEATHER, POVERTY-GRASS.

Lechea intermedia Legg. ex Britton   ROUND-FRUITED PINWEED.

Lechea leggettii Britton & Hollick   LEGGETT'S PINWEED.   (Incl. var. *moniliformis* (E. P. Bicknell) Hodgdon; treated by some authors as L. *pulchella* Raf.)

Lechea minor L.   THYME-LEAVED PINWEED. (Treated by some authors as L. *thymifolia* Michx.)

Lechea racemulosa Michx.   OBLONG-FRUITED PINWEED.

Lechea tenuifolia Michx.   NARROW-LEAVED PINWEED.

Lechea villosa Elliott   HAIRY PINWEED. (Treated by some authors as L. *mucronata* Raf.)

■ VIOLACEAE.   VIOLET FAMILY

Hybanthus concolor (T. F. Forst.) Spreng.   GREEN VIOLET.

Viola affinis Leconte   THIN-LEAVED VIOLET, LECONTE'S VIOLET.   (V. *sororia* var. *affinis* (Leconte) L. E. McKinney)

***Viola arvensis** Murray   EUROPEAN FIELD PANSY, EUROPEAN WILD PANSY.

Viola blanda Willd.   SWEET WHITE VIOLET. (Incl. var. *palustriformis* A. Gray and V. *incognita* Brainerd)

Viola canadensis L.   CANADA VIOLET.

Viola conspersa Rchb.   AMERICAN DOG VIOLET.

Viola cucullata Aiton   MARSH BLUE VIOLET. (Treated by some authors as V. *obliqua* Hill)

Viola fimbriatula Sm.   OVATE-LEAVED VIOLET, NORTHERN DOWNY VIOLET.   (V. *sagittata* var. *ovata* (Nutt.) Torr. & A. Gray)

Viola hastata Michx.   HALBERD-LEAVED VIOLET.

Viola hirsutula Brainerd   SOUTHERN WOOD VIOLET.   (Treated by some authors as a part of V. *villosa* Walter)

Viola lanceolata L.   LANCE-LEAVED VIOLET.

Viola macloskeyi F. E. Lloyd **var. pallens** (Banks ex DC.) C. L. Hitchc.   NORTHERN WHITE VIOLET.   (subsp. *pallens* (Banks ex DC.) M. S. Baker, V. *pallens* (Banks ex DC.) Brainerd)

Viola missouriensis Greene   MISSOURI VIOLET. (V. *sororia* var. *missouriensis* (Greene) L. E. McKinney)

Viola nephrophylla Greene   NORTHERN BOG VIOLET.   (Treated by some authors as a part of V. *sororia* var. *sororia*)

***Viola odorata** L.   SWEET VIOLET.

Viola palmata L.   EARLY BLUE VIOLET, PALMATE-LEAVED VIOLET.

Viola pedata L.   BIRDFOOT VIOLET.

Viola pedatifida G. Don   PRAIRIE VIOLET,

Larkspur Violet. (*V. palmata* var. *pedatifida* (G. Don) Cronquist)

Viola primulifolia L. Primrose-leaved Violet.

Viola pubescens Aiton Common Yellow Violet.

    var. pubescens Downy Common Yellow Violet. (Incl. *V. pensylvanica* Michx.)

    var. scabriuscula Torr. & A. Gray Smooth Common Yellow Violet. (Incl. *V. eriocarpa* Schwein. and *V. pensylvanica* of some authors).

Viola rafinesquii Greene Wild Pansy, Field Pansy. (Treated by some authors as *V. bicolor* Pursh)

Viola rostrata Pursh Long-spurred Violet.

Viola rotundifolia Michx. Round-leaved Violet, Early Yellow Violet.

Viola sagittata Aiton Arrow-leaved Violet.

Viola sororia Willd. **var. sororia** (Incl. *V. papilionacea* Pursh) Common Blue Violet.

    forma beckwithiae House Beckwith's Violet.

    forma priceana (Pollard) Cooperr. Confederate Violet. (*V. priceana* Pollard, *V. papilionacea* var. *priceana* (Pollard) Alexander)

Viola striata Aiton Striped Violet, Common White Violet.

*****Viola tricolor** L. Johnny-jump-up.

Viola triloba Schwein. Three-lobed Violet. (*V. palmata* var. *triloba* (Schwein.) Ging. ex DC.)

Viola tripartita Elliott

    var. glaberrima (DC.) R. M. Harper Southern Appalachian Yellow Violet, Wedge-leaved Violet. (*V. tripartita* forma *glaberrima* (DC.) Fernald)

    var. tripartita Three-parted Violet. (*V. tripartita* forma *tripartita*)

Viola walteri House Walter's Violet.

Viola × aberrans Greene (**Viola fimbriatula × V. sororia**) Aberrant Violet. (Treated by some authors as a part of *V. × conjugens* Greene)

Viola × bissellii House (**Viola cucullata × V. sororia**) Bissell's Violet.

Viola × brauniae Grover ex Cooperr. (**Viola rostrata × V. striata**) Braun's Violet.

Viola × cooperrideri H. E. Ballard (**Viola striata × V. walteri**) Cooperrider's Violet.

Viola × dissita House (**Viola hirsutula × V. triloba**) House's Violet. (Treated by some authors as a part of *V. ravida* House)

Viola × emarginata (Nutt.) Leconte (**Viola affinis × V. sagittata**) Triangle-leaved Violet.

Viola × festata House (**Viola cucullata × V. sagittata**) Festive Violet.

Viola × filicetorum Greene (**Viola affinis × V. sororia**) Fernery Violet.

Viola × malteana House (**Viola conspersa × V. rostrata**) Malte's Violet.

Viola × milleri Moldenke (**Viola affinis × V. triloba**) Miller's Violet. (Treated by some authors as a part of *V. slavinii* House)

Viola × populifolia Greene (**Viola sororia × V. triloba**) Poplar-leaved Violet.

Viola × porteriana Pollard (**Viola cucullata × V. fimbriatula**) Porter's Violet.

†**Viola × wittrockiana** Gams (**Viola tricolor ×** other undetermined parent) Garden Pansy.

■ TAMARICACEAE. Tamarisk Family

†**Tamarix chinensis** Lour. Chinese Tamarisk.

■ PASSIFLORACEAE. Passion-flower Family

Passiflora incarnata L. Maypop, Purple Passion-flower.

Passiflora lutea L. Yellow Passion-flower. (Incl. var. *glabriflora* Fernald)

■ CUCURBITACEAE. Cucumber Family or Gourd Family

†**Citrullus lanatus** (Thunb.) Matsum. & Nakai Watermelon.

†**Cucumis melo** L. Muskmelon.

†**Cucurbita foetidissima** Kunth Missouri Gourd, Wild Pumpkin.

†**Cucurbita maxima** Duchesne   SQUASH.
†**Cucurbita pepo** L.   PUMPKIN.
Echinocystis lobata (Michx.) Torr. & A. Gray
   WILD CUCUMBER, BALSAM-APPLE.
Sicyos angulatus L.   BUR-CUCUMBER, STAR-
   CUCUMBER.

Order **Salicales**

■ SALICACEAE.   WILLOW FAMILY

*****Populus alba** L.   WHITE POPLAR, SILVER
   POPLAR.
Populus balsamifera L.   BALSAM POPLAR.
   (Incl. var. *subcordata* Hyl.)
Populus deltoides W. Bartram ex Marshall
   COTTONWOOD.
   var. deltoides   SOUTHERN COTTONWOOD.
      (subsp. *deltoides*; incl. var. *missouriensis*
      (A. Henry) A. Henry)
   var. occidentalis Rydb.   NORTHERN
      COTTONWOOD.   (subsp. *monilifera* (Aiton)
      Eckenw.)
Populus grandidentata Michx.   BIGTOOTH
   ASPEN.
Populus heterophylla L.   SWAMP COTTONWOOD.
†**Populus nigra** L. **var. italica** Münchh.
   LOMBARDY POPLAR, BLACK POPLAR.   (Plants
   of this variety are treated instead by some
   authors as cultivar 'Italica'.)
Populus tremuloides Michx.   QUAKING ASPEN.
†**Populus × canadensis** Moench (**Populus del-**
   toides × P. nigra)   CAROLINA POPLAR.
†**Populus × canescens** (Aiton) Sm. (**Populus**
   alba × P. tremula L.)   GRAY POPLAR.
Populus × jackii Sarg. (**Populus balsamifera ×**
   P. deltoides)   BALM-OF-GILEAD.   (Incl.
   P. gileadensis Rouleau)
Populus × smithii B. Boivin (**Populus grandi-**
   dentata × P. tremuloides)   SMITH'S POPLAR.
   (Incl. *P. barnesii* W. H. Wagner)
*****Salix alba** L.
   *****var. alba**   WHITE WILLOW.
   *****var. calva** G. Mey.   CRICKET-BAT WILLOW.
   *****var. vitellina** (L.) Stokes   GOLDEN WILLOW.
Salix amygdaloides Andersson   PEACH-LEAVED
   WILLOW.

†**Salix babylonica** L.   WEEPING WILLOW.
Salix bebbiana Sarg.   BEAKED WILLOW, BEBB'S
   WILLOW.
Salix candida Flüggé ex Willd.   SAGE-LEAVED
   WILLOW, HOARY WILLOW.
†**Salix caprea** L.   GOAT WILLOW, FLORIST'S
   WILLOW.
Salix caroliniana Michx.   CAROLINA WILLOW,
   WARD'S WILLOW.
†**Salix cinerea** L.   GRAY FLORIST'S WILLOW.
Salix discolor Muhl.   PUSSY WILLOW.   (Incl.
   var. *latifolia* Andersson)
Salix eriocephala Michx.   HEART-LEAVED
   WILLOW, DIAMOND WILLOW.   (Treated by
   some authors as *S. rigida* Muhl., incl. var.
   angustata (Pursh) Fernald)
*****Salix fragilis** L.   CRACK WILLOW, BRITTLE
   WILLOW.
Salix humilis Marshall
   var. humilis   PRAIRIE WILLOW, UPLAND
      WILLOW.   (Incl. var. *hyporhysa* Fernald
      and var. *keweenawensis* Farw.)
   var. tristis (Aiton) Griggs   DWARF UPLAND
      WILLOW.   (Incl. var. *microphylla* (Anders-
      son) Fernald; *S. occidentalis* Walter; *S. tristis*
      Aiton)
Salix interior Rowlee   SANDBAR WILLOW.
   (Treated by some authors as a part of *S. exigua*
   Nutt.)
   var. angustissima (Andersson) Dayton (Incl.
      var. *wheeleri* Rowlee; *S. exigua* Nutt. var.
      angustissima (Andersson) Reveal & C. R.
      Broome)
   var. pedicellata (Andersson) C. R. Ball
      (*S. exigua* Nutt. var. *pedicellata* (Andersson)
      Cronquist)
Salix lucida Muhl.   SHINING WILLOW.
†**Salix matsudana** Koidz.   CORKSCREW WILLOW,
   PEKIN WILLOW.
Salix myricoides Muhl.   BLUE-LEAVED WILLOW,
   DUNE WILLOW.   (Incl. *S. glaucophylloides*
   Fernald)
   var. albovestita (C. R. Ball) Dorn
   var. myricoides
Salix nigra Marshall   BLACK WILLOW.
Salix pedicellaris Pursh   BOG WILLOW.   (Incl.
   var. *hypoglauca* Fernald)

*Salix pentandra L.   Bay-leaved Willow.

Salix petiolaris Sm.   Meadow Willow, Slender Willow.   (Incl. Ohio plants that have in the past been identified as *S. subsericea* (Andersson) C. K. Schneid.)

*Salix purpurea L.   Basket Willow, Purple Osier.

Salix sericea Marshall   Silky Willow.

Salix serissima (L. H. Bailey) Fernald   Autumn Willow.

†Salix viminalis L.   Silky Osier, Basket Willow.

Salix × bebbii Gand. (Salix eriocephala × S. sericea)   Gandoger's Willow.

Salix × glatfelteri C. K. Schneid. (Salix amygda-loides × S. nigra)   Glatfelter's Willow.

*Salix × jesupii Fernald (Salix alba × S. lucida)   Jesup's Willow.

*Salix × rubens Schrank (Salix alba × S. fragilis)   Schrank's Willow.

Salix eriocephala × S. petiolaris

Order **Capparales**

■ CAPPARACEAE.   Caper Family

†Cleome hassleriana Chodat   Spider-flower.

†Cleome serrulata Pursh   Stinking-clover, Rocky Mountain Bee-plant.

Polanisia dodecandra (L.) DC.   Clammy-weed.

†Polanisia jamesii (Torr. & A. Gray) H. H. Iltis   James's Clammy-weed.   (*Cristatella jamesii* Torr. & A. Gray)

■ BRASSICACEAE or CRUCIFERAE.   Mustard Family

*Alliaria petiolata (M. Bieb.) Cavara & Grande   Garlic Mustard.

†Alyssum alyssoides (L.) L.   Alyssum.

†Alyssum saxatile L.   Yellowtuft, Goldentuft Madwort, Basket-of-gold.   (*Aurinia saxatilis* (L.) Desv.)

*Arabidopsis thaliana (L.) Heynh.   Mouse-ear Cress.

Arabis canadensis L.   Sickle-pod, Sickle-pod Rock Cress.

Arabis divaricarpa A. Nelson   Limestone Rock Cress.

Arabis drummondii A. Gray   Drummond's Rock Cress.

Arabis glabra (L.) Bernh.   Tower Mustard.

Arabis hirsuta (L.) Scop.   Hairy Rock Cress.
  var. adpressipilis (M. Hopkins) Rollins   Southern Hairy Rock Cress.
  var. pycnocarpa (M. Hopkins) Rollins   Western Hairy Rock Cress.

Arabis laevigata (Muhl. ex Willd.) Poir.   Smooth Rock Cress.

Arabis lyrata L.   Sand Cress, Lyre-leaved Rock Cress.

Arabis patens Sull.   Spreading Rock Cress.

Arabis shortii (Fernald) Gleason.   Short's Rock Cress.   (*A. perstellata* E. L. Braun var. *shortii* Fernald)

Armoracia lacustris (A. Gray) Al-Shehbaz & V. M. Bates   Lake Cress.   (Treated by some authors as *A. aquatica* (Eaton) Britton or *Neobeckia aquatica* (Eaton) Greene)

†Armoracia rusticana P. Gaertn., B. Mey. & Scherb.   Horseradish.

*Barbarea verna (Mill.) Asch.   Early Winter Cress.

*Barbarea vulgaris R. Br.   Yellow Rocket, Common Winter Cress.

*Berteroa incana (L.) DC.   Hoary-alyssum.

*Brassica juncea (L.) Czern.   Brown Mustard, Indian Mustard.

†Brassica napus L.   Rape, Rutabaga.   (Incl. *B. napobrassica* (L.) Mill.)

*Brassica nigra (L.) K. Koch   Black Mustard.

*Brassica oleracea L.   Cabbage.

*Brassica rapa L.   Field Mustard, Turnip.

†Bunias orientalis L.   Turkish Rocket, Hill Mustard.

Cakile edentula (Bigelow) Hook. var. lacustris Fernald   Inland Sea Rocket.   (subsp. *lacustris* (Fernald) Hultén)

*Camelina microcarpa Andrz. ex DC.   Small-seeded False Flax.

†Camelina sativa (L.) Crantz   Large-seeded False Flax.

*Capsella bursa-pastoris (L.) Medik.   Shepherd's-purse.

Cardamine angustata O. E. Schulz
APPALACHIAN TOOTHWORT, SLENDER
TOOTHWORT. (*Dentaria heterophylla* Nutt.)

Cardamine bulbosa (Schreb. ex Muhl.) Britton,
Sterns & Poggenb. BULBOUS BITTER
CRESS, SPRING CRESS. (Incl. *C. rhomboidea*
(Pers.) DC.)

Cardamine concatenata (Michx.) Sw.
CUT-LEAVED TOOTHWORT, FIVE-PARTED
TOOTHWORT. (*Dentaria laciniata* Muhl.
ex Willd.)

Cardamine diphylla (Michx.) A. W. Wood
TWO-LEAVED TOOTHWORT, BROAD-LEAVED
TOOTHWORT. (*Dentaria diphylla* Michx.)

Cardamine dissecta (Leavenw.) Al-Shehbaz
NARROW-LEAVED TOOTHWORT. (*Dentaria
multifida* Muhl. ex Elliott)

Cardamine douglassii Britton PURPLE BITTER
CRESS, PINK SPRING CRESS.

*Cardamine flexuosa With. BENDING BITTER
CRESS.

*Cardamine hirsuta L. HOARY BITTER CRESS.

† Cardamine impatiens L. EUROPEAN BITTER
CRESS.

Cardamine parviflora L. **var. arenicola** (Britton)
O. E. Schulz DRY-LAND BITTER CRESS.

Cardamine pensylvanica Muhl. ex Willd.
PENNSYLVANIA BITTER CRESS.

Cardamine pratensis L. CUCKOO-FLOWER.
var. palustris Wimm. & Grab. AMERICAN
CUCKOO-FLOWER.
† **var. pratensis.** EURASIAN CUCKOO-FLOWER.

Cardamine rotundifolia Michx. TRAILING
BITTER CRESS, ROUND-LEAVED BITTER CRESS.

*Cardaria draba (L.) Desv. HOARY CRESS.

† Chorispora tenella (Pall.) DC. BLUE
MUSTARD.

*Conringia orientalis (L.) Andrz. HARE'S-EAR
MUSTARD.

*Coronopus didymus (L.) Sm. SWINE CRESS,
LESSER WART CRESS.

Descurainia pinnata (Walter) Britton **var.
brachycarpa** (Richardson) Fernald TANSY
MUSTARD. (subsp. *brachycarpa* (Richardson)
Detling)

*Descurainia sophia (L.) Webb ex Prantl HERB
SOPHIA.

*Diplotaxis muralis (L.) DC. SAND ROCKET,
STINKING WALL ROCKET.

*Diplotaxis tenuifolia (L.) DC. SLIM-LEAVED
WALL ROCKET.

Draba brachycarpa Nutt. ex Torr. & A. Gray
LITTLE WHITLOW-GRASS.

Draba cuneifolia Nutt. ex Torr. & A. Gray
WEDGE-LEAVED WHITLOW-GRASS.

Draba reptans (Lam.) Fernald CAROLINA
WHITLOW-GRASS.

*Erophila verna (L.) Besser VERNAL WHITLOW-
GRASS. (*Draba verna* L.)

*Erucastrum gallicum (Willd.) O. E. Schulz
DOG MUSTARD.

Erysimum capitatum (Douglas ex Hook.) Greene
WESTERN WALLFLOWER. (Incl. *E. arkan-
sanum* Nutt.)

*Erysimum cheiranthoides L. WORMSEED
MUSTARD.

† Erysimum inconspicuum (S. Watson) MacMill.
PLAINS WALLFLOWER.

† Erysimum repandum L. TREACLE MUSTARD,
BUSHY WALLFLOWER.

*Hesperis matronalis L. DAME'S ROCKET,
SWEET ROCKET.

† Iberis umbellata L. GLOBE CANDYTUFT.

Iodanthus pinnatifidus (Michx.) Steud.
PURPLE ROCKET.

Leavenworthia uniflora (Michx.) Britton
MICHAUX'S LEAVENWORTHIA.

*Lepidium campestre (L.) R. Br. FIELD PEPPER-
GRASS, COW CRESS.

Lepidium densiflorum Schrad. WILD PEPPER-
GRASS, PRAIRIE PEPPERWEED.

*Lepidium perfoliatum L. CLASPING PEPPER-
GRASS, CLASPING PEPPERWEED.

† Lepidium ramosissimum A. Nelson BUSHY
PEPPER-GRASS.

*Lepidium ruderale L. STINKING PEPPERWEED,
STINKING PEPPER-GRASS.

† Lepidium sativum L. GARDEN CRESS.

Lepidium virginicum L. POOR-MAN'S-PEPPER,
VIRGINIA PEPPER-GRASS.

† Lobularia maritima (L.) Desv. SWEET
ALYSSUM.

*Lunaria annua L. HONESTY, SILVER-DOLLAR,
MONEY-PLANT.

*Neslia paniculata (L.) Desv.   Ball Mustard.

*Raphanus raphanistrum L.   Wild Radish, Jointed Charlock.

†Raphanus sativus L.   Radish.

*Rorippa nasturtium-aquaticum (L.) Hayek Watercress.   (*Nasturtium officinale* W. T. Aiton)

Rorippa palustris (L.) Besser   Common Yellow Cress.
  var. fernaldiana (Butters & Abbe) Stuckey (subsp. *fernaldiana* (Butters & Abbe) Jonsell)
  var. hispida (Desv.) Rydb. (subsp. *hispida* (Desv.) Jonsell)

Rorippa sessiliflora (Nutt.) Hitchc.   Southern Yellow Cress.

*Rorippa sylvestris (L.) Besser   Creeping Yellow Cress.

*Sibara virginica (L.) Rollins   Virginia Rock Cress.

*Sinapis alba L.   White Mustard.   (*Brassica alba* (L.) Rabenh.)

*Sinapis arvensis L.   Charlock, Wild Mustard.   (*Brassica kaber* (DC.) L. C. Wheeler)

*Sisymbrium altissimum L.   Tumbling Mustard.

†Sisymbrium irio L.   London Rocket.

†Sisymbrium loeselii L.   Tall Hedge Mustard.

*Sisymbrium officinale (L.) Scop.   Hedge Mustard.

†Thlaspi alliaceum L.   Garlic Penny Cress.

*Thlaspi arvense L.   Field Penny Cress.

*Thlaspi perfoliatum L.   Perfoliate Penny Cress, Thoroughwort Penny Cress.

■ RESEDACEAE.   Mignonette Family

†Reseda alba L.   White Mignonette.

†Reseda lutea L.   Yellow Mignonette.

†Reseda luteola L.   Dyer's-rocket.

Order Ericales

■ ERICACEAE.   Heath Family

Andromeda glaucophylla Link   Bog-rosemary.   (*A. polifolia* L. var. *glaucophylla* (Link) DC.)

Arctostaphylos uva-ursi (L.) Spreng.   Bearberry, Kinnikinick.   (Incl. var. *coactilis* Fernald & J. F. Macbr.)

*Calluna vulgaris (L.) Hull   Heather.

Chamaedaphne calyculata (L.) Moench Leather-leaf.   (Incl. var. *angustifolia* (Aiton) Rehder)

Epigaea repens L.   Trailing Arbutus, Mayflower.   (Incl. var. *glabrifolia* Fernald)

†Erica tetralix L.   Cross-leaved Heath.

Gaultheria hispidula (L.) Muhl. ex Bigelow Creeping Snowberry.

Gaultheria procumbens L.   Wintergreen, Checkerberry, Teaberry.

Gaylussacia baccata (Wangenh.) Koch Huckleberry, Black Huckleberry.

Kalmia latifolia L.   Mountain Laurel.

Ledum groenlandicum Oeder   Labrador-tea.

†Leucothoe recurva (Buckley) A. Gray   Red-twig.

Lyonia ligustrina (L.) DC.   Male-berry.

Oxydendrum arboreum (L.) DC.   Sourwood, Sorrel-tree.

Rhododendron calendulaceum (Michx.) Torr.   Flame Azalea.

Rhododendron maximum L.   Great Rhododendron, Rosebay Rhododendron.

Rhododendron nudiflorum (L.) Torr.
  var. nudiflorum   Pinxter-flower.   (Incl. *R. periclymenoides* (Michx.) Shinners)
  var. roseum (Loisel.) Wiegand   Roseshell Azalea, Northern Rose Azalea.   (*R. roseum* (Loisel.) Rehder; incl. *R. prinophyllum* (Small) Millais)

Vaccinium angustifolium Aiton   Low Sugarberry, Sweet Lowbush Blueberry, Late Lowbush Blueberry.   (Incl. *V. brittonii* Porter ex E. P. Bicknell and *V. lamarckii* Camp)

Vaccinium corymbosum L.   Highbush Blueberry.   (Incl. *V. atrococcum* (A. Gray) A. Heller and *V. simulatum* Small)

Vaccinium macrocarpon Aiton   Large Cranberry.

Vaccinium myrtilloides Michx.   Velvet-leaved Blueberry.

Vaccinium oxycoccos L.   Small Cranberry.

Vaccinium pallidum Aiton Low Blueberry.
(Incl. *V. altomontanum* Ashe and *V. vacillans*
Kalm ex Torr.)

Vaccinium stamineum L. Deerberry, Squaw-
huckleberry. (Incl. var. *melanocarpum*
C. Mohr, var. *neglectum* (Small) Deam, and
V. caesium Greene)

■ PYROLACEAE. Pyrola Family or
Shinleaf Family

Chimaphila maculata (L.) Pursh Spotted
Pipsissewa, Spotted Wintergreen.

Chimaphila umbellata (L.) W. P. C. Barton **var.
cisatlantica** S. F. Blake Pipsissewa, Prince's-
pine. (subsp. *cisatlantica* (S. F. Blake) Hultén)

Moneses uniflora (L.) A. Gray One-flowered
Wintergreen.

Orthilia secunda (L.) House One-sided
Wintergreen. (*Pyrola secunda* L.)

Pyrola chlorantha Sw. Green-flowered
Shinleaf, Green-flowered Wintergreen.

Pyrola elliptica Nutt. Shinleaf.

Pyrola rotundifolia L. **var. americana** (Sweet)
Fernald Round-leaved Wintergreen,
Wild Lily-of-the-valley. (subsp. *americana*
(Sweet) R. T. Clausen, *P. americana* Sweet,
P. asarifolia Michx. subsp. *americana* (Sweet)
Krísa)

■ MONOTROPACEAE. Indian Pipe Family

Monotropa hypopithys L. Pinesap.

Monotropa uniflora L. Indian Pipe.

Order **Diapensiales**

■ DIAPENSIACEAE. Diapensia Family

†**Galax urceolata** (Poir.)Brummitt
Wandflower. (Treated by some authors
as a part of G. *aphylla* L.)

Order **Ebenales**

■ EBENACEAE. Ebony Family

Diospyros virginiana L. Persimmon.

■ STYRACACEAE. Storax Family

Halesia carolina L. Silver-bell, Carolina
Silver-bell. (Incl. *H. tetraptera* Ellis)

Styrax americanus Lam. Snowbell.

Styrax grandifolius Aiton Bigleaf
Snowbell.

Order **Primulales**

■ PRIMULACEAE. Primrose Family

*****Anagallis arvensis** L. Scarlet Pimpernel.

Androsace occidentalis Pursh Western Rock-
jasmine.

Centunculus minimus L. Chaffweed.
(*Anagallis minima* (L.) Krause)

Dodecatheon meadia L. Shooting-star,
Pride-of-Ohio.

Hottonia inflata Elliott Featherfoil, Water-
violet.

Lysimachia ciliata L. Fringed Loosestrife.
(*Steironema ciliatum* (L.) Baudo)

Lysimachia lanceolata Walter Lance-leaved
Loosestrife. (*Steironema lanceolatum*
(Walter) A. Gray)

*****Lysimachia nummularia** L. Moneywort.

*****Lysimachia punctata** L. Garden
Loosestrife.

Lysimachia quadriflora Sims Linear-leaved
Loosestrife. (*Steironema quadriflorum*
(Sims) Hitchc.)

Lysimachia quadrifolia L. Whorled
Loosestrife.

Lysimachia terrestris (L.) Britton, Sterns &
Poggenb. Swamp-candles.

Lysimachia thyrsiflora L. Tufted
Loosestrife.

*****Lysimachia vulgaris** L. Garden
Loosestrife.

Lysimachia × producta (A. Gray) Fernald
(**Lysimachia quadrifolia × L. terrestris**)
Gray's Loosestrife.

Samolus parviflorus Raf. Water-pimpernel.
(Incl. *S. floribundus* Kunth; *S. valerandi* L.
subsp. *parviflorus* (Raf.) Hultén)

Trientalis borealis Raf. Starflower.

Subclass **ROSIDAE**

Order **Rosales**

■ HYDRANGEACEAE. Hydrangea Family

Hydrangea arborescens L. Wild Hydrangea, Hills-of-snow. (Incl. var. *oblonga* Torr. & A. Gray)

†**Philadelphus coronarius** L. European Mock Orange.

†**Philadelphus inodorus** L. Appalachian Mock Orange.

*****Philadelphus tomentosus** Wall. Downy Mock Orange.

■ GROSSULARIACEAE. Gooseberry Family or Currant Family

Ribes americanum Mill. Wild Black Currant, American Black Currant.

Ribes cynosbati L. Prickly Gooseberry, Dogberry. (Incl. var. *glabrum* Fernald)

Ribes glandulosum Grauer Skunk Currant.

Ribes hirtellum Michx. Smooth Gooseberry.
  var. calcicola (Fernald) Fernald
  var. hirtellum

Ribes missouriense Nutt. Missouri Gooseberry.

†**Ribes nigrum** L. Garden Black Currant.

†**Ribes odoratum** H. Wendl. Buffalo Currant. (*R. aureum* Pursh var. *villosum* DC.)

†**Ribes oxyacanthoides** L. Northern Gooseberry, Hawthorn-leaved Gooseberry.

†**Ribes rubrum** L. Garden Red Currant. (Incl. *R. sativum* Syme)

Ribes triste Pall. Swamp Red Currant.

†**Ribes uva-crispa** L. **var. sativum** DC. Garden Gooseberry, English Gooseberry. (*R. grossularia* L.)

■ CRASSULACEAE. Stonecrop Family

*****Sedum acre** L. Golden-carpet, Mossy Stonecrop.

†**Sedum album** L. White Stonecrop.

†**Sedum dendroideum** Moç. & Sessé ex A. DC. Shrubby Stonecrop.

*****Sedum sarmentosum** Bunge Creeping Stonecrop.

†**Sedum sexangulare** L. Six-ranked Stonecrop.

†**Sedum telephioides** Michx. American Orpine.

*****Sedum telephium** L. Garden Orpine, Live-forever. (Incl. var. *purpureum* L.)

Sedum ternatum Michx. Three-branched Stonecrop, Wild Stonecrop.

■ SAXIFRAGACEAE. Saxifrage Family

Chrysosplenium americanum Schwein. ex Hook. Water-carpet, Golden-saxifrage.

Heuchera americana L. Common Alum-root, Rock-geranium.
  var. americana
  var. hirsuticaulis (Wheelock) Rosend., Butters, & Lakela
  var. hispida (Pursh) E. Wells

Heuchera longiflora Rydb. Long-flowered Alum-root, Close-flowered Alum-root.

Heuchera parviflora Bartl. **var. rugelii** (Shuttlew.) Rosend., Butters & Lakela Small-flowered Alum-root.

Heuchera villosa Michx. Hairy Alum-root, Maple-leaved Alum-root.

Mitella diphylla L. Bishop's-cap, Mitrewort —also spelled Miterwort.

Parnassia glauca Raf. Grass-of-Parnassus.

Penthorum sedoides L. Ditch-stonecrop.

Saxifraga pensylvanica L. Swamp Saxifrage.

Saxifraga virginiensis Michx. Early Saxifrage.

Sullivantia sullivantii (Torr. & A. Gray) Britton Sullivantia.

Tiarella cordifolia L. Foamflower.

■ ROSACEAE. Rose Family

†**Agrimonia eupatoria** L. Medicinal Agrimony.

Agrimonia gryposepala Wallr.   COMMON
AGRIMONY, HAIRY AGRIMONY.

†**Agrimonia microcarpa** Wallr.   LOW
AGRIMONY.

Agrimonia parviflora Aiton   SOUTHERN
AGRIMONY, SMALL-FLOWERED AGRIMONY.

Agrimonia pubescens Wallr.   DOWNY
AGRIMONY, SOFT AGRIMONY.

Agrimonia rostellata Wallr.   WOODLAND
AGRIMONY.

Agrimonia striata Michx.   ROADSIDE
AGRIMONY, STRIATE AGRIMONY.

Amelanchier arborea (F. Michx.) Fernald
DOWNY SERVICEBERRY.

Amelanchier interior Nielsen
INLAND SERVICEBERRY.

Amelanchier laevis Wiegand   SMOOTH
SERVICEBERRY, ALLEGHENY SERVICEBERRY.

Amelanchier sanguinea (Pursh) DC.   ROCK
SERVICEBERRY, NEW ENGLAND SERVICEBERRY.

Amelanchier spicata (Lam.) K. Koch   DWARF
SERVICEBERRY.   (Incl. *A. stolonifera* Wiegand)

Amelanchier arborea × A. laevis

Amelanchier arborea × A. sanguinea

Aruncus dioicus (Walter) Fernald   GOAT'S-
BEARD.

†**Chaenomeles speciosa** (Sweet) Nakai
JAPANESE-QUINCE, FLOWERING-QUINCE.
(Treated by some authors as *C. lagenaria*
(Loisel.) Koidz.)

†**Cotoneaster divaricatus** Rehder & E. H. Wilson
SPREADING COTONEASTER.

†**Cotoneaster simonsii** Baker   SIMONS'
COTONEASTER.

Crataegus brainerdii Sarg.   BRAINERD'S
HAWTHORN.   (Incl. var. *brainerdii*, var.
scabrida (Sarg.) Eggl., and *C. coleae* Sarg.)

Crataegus calpodendron (Ehrh.) Medik.   PEAR
HAWTHORN, BLACK THORN.   (Incl. var. *calpodendron*, var. *globosa* (Sarg.) E. J. Palmer, and
var. *microcarpa* (Chapm.) E. J. Palmer)

Crataegus chrysocarpa Ashe   FIREBERRY
HAWTHORN.   (Incl. *C. margarettiae* var. *margarettiae*, var. *brownii* (Britton) Sarg., and var.
meiophylla (Sarg.) E. J. Palmer, and *C. sicca*
Sarg. var. *glabrifolia* (Sarg.) E. J. Palmer; treated
by some authors as *C. rotundifolia* Lam.)

Crataegus coccinea L.   SCARLET HAWTHORN.
(Incl. *C. habereri* Sarg., *C. hillii* Sarg., *C. holme-
siana* Ashe, *C. pennsylvanica* Ashe, *C. pringlei*
Sarg., *C. putnamiana* Sarg., and *C. pedicellata*
Sarg. var. *pedicellata*, var. *albicans* (Ashe) E. J.
Palmer, var. *assurgens* (Sarg.) E. J. Palmer, and
var. *robesoniana* (Sarg.) E. J. Palmer)

Crataegus crus-galli L.   COCKSPUR, COCKSPUR
THORN.   (Incl. var. *crus-galli*, var. *barrettiana*
(Sarg.) E. J. Palmer, var. *exigua* (Sarg.) Eggl.,
var. *leptophylla* (Sarg.) E. J. Palmer, var. *pachy-
phylla* (Sarg.) E. J. Palmer, and var. *pyracan-
thifolia* Aiton; and incl. *C. arborea* Beadle
(*C. pyracanthoides* (Aiton) Beadle var. *arborea*
(Beadle) E. J. Palmer), *C. engelmannii* Sarg.,
C. fontanesiana of authors, *C. hannibalensis*
E. J. Palmer, *C. ohioensis* Sarg., and *C. vallicola*
Sarg.)

Crataegus flabellata (Spach) G. Kirchn.   FAN-
LEAVED HAWTHORN.   (Incl. *C. basilica* Beadle,
C. beata Sarg., *C. brumalis* Ashe, *C. gravis*
Ashe, *C. iracunda* Beadle var. *silvicola* (Beadle)
E. J. Palmer, *C. macrosperma* Ashe var. *macro-
sperma*, var. *acutiloba* (Sarg.) Eggl., var. *demissa*
(Sarg.) Eggl., var. *matura* (Sarg.) Eggl., var.
pentranda (Sarg.) Eggl., and var. *roanensis*
(Ashe) E. J. Palmer, *C. populnea* Ashe, and
C. stolonifera Sarg.)

Crataegus intricata Lange   BILTMORE
HAWTHORN.   (Incl. var. *intricata* and var.
straminea (Beadle) E. J. Palmer, *C. biltmoreana*
Beadle, *C. boyntonii* Beadle, *C. fortunata*
Sarg., *C. horseyi* E. J. Palmer, and *C. rubella*
Beadle)

†**Crataegus laevigata** (Poir.) DC.   ENGLISH
HAWTHORN.   (Incl. plants formerly called
by the now rejected name *C. oxyacantha* L.)

Crataegus mollis Scheele   DOWNY HAWTHORN.
(Incl. var. *mollis* and var. *sera* (Sarg.) Eggl.,
and *C. submollis* Sarg.)

*****Crataegus monogyna** Jacq.   ONE-SEEDED
HAWTHORN.

Crataegus phaenopyrum (L.f.) Medik.
WASHINGTON THORN, WASHINGTON
HAWTHORN.

Crataegus pruinosa (H. L. Wendl.) K. Koch
WAXY-FRUITED HAWTHORN, FROSTED

Hawthorn. (Incl. var. *pruinosa*, var. *dissona* (Sarg.) Eggl., var. *latisepala* (Sarg.) Eggl., C. *compacta* Sarg., C. *crawfordiana* Sarg., C. *disjuncta* Sarg., C. *formosa* Sarg., C. *franklinensis* Sarg., C. *gattingeri* Ashe, C. *gaudens* Sarg., C. *jesupii* Sarg., C. *leiophylla* Sarg., C. *mackenzii* Sarg. var. *bracteata* (Sarg.) E. J. Palmer, C. *milleri* Sarg., C. *porteri* Britton, C. *rugosa* Ashe, and C. *virella* Ashe)

Crataegus punctata Jacq. DOTTED HAWTHORN. (Incl. var. *punctata*, var. *aurea* Aiton, var. *canescens* Britton, var. *microphylla* Sarg., var. *pausiaca* (Ashe) E. J. Palmer, C. *indicens* Ashe, C. *peoriensis* Sarg., and C. *suborbiculata* Sarg.)

Crataegus succulenta Schrad. ex Link LONG-THORNED HAWTHORN, FLESHY HAWTHORN. (Incl. var. *succulenta*, var. *macrantha* (Lodd.) Eggl., var. *michiganensis* (Ashe) E. J. Palmer, var. *neofluvialis* (Ashe) E. J. Palmer, and var. *pertomentosa* (Ashe) E. J. Palmer)

Crataegus uniflora Münchh. DWARF HAWTHORN, ONE-FLOWERED HAWTHORN.

Crataegus × anomala Sarg. (**Crataegus intricata × C. mollis**) SARGENT'S HAWTHORN.

Crataegus × chadsfordiana Sarg. (**Crataegus pruinosa × C. succulenta**) CHADSFORD HAWTHORN.

Crataegus × disperma Ashe (**Crataegus crusgalli × C. punctata**) TWO-SEEDED HAWTHORN.

Crataegus × hudsonica Sarg. (**Crataegus pruinosa × C. punctata**) HUDSON'S HAWTHORN. (Incl. C. × *kellermanii* Sarg.)

Crataegus × locuples Sarg. (**Crataegus mollis × C. pruinosa**) RICH HAWTHORN.

Crataegus × lucorum Sarg. (**Crataegus coccinea × C. flabellata**) WOODLAND HAWTHORN.

Crataegus × persimilis Sarg. (**Crataegus crusgalli × C. succulenta**) NEW YORK HAWTHORN. (Incl. C. × *laetifica* Sarg.)

Crataegus calpodendron × C. crus-galli

Crataegus coccinea × C. pruinosa

† **Cydonia oblonga** Mill. COMMON QUINCE.

Dalibarda repens L. ROBIN-RUN-AWAY, DEWDROP.

* **Duchesnea indica** (Andrews) Focke INDIAN-STRAWBERRY.

Filipendula rubra (Hill) B. L. Rob. QUEEN-OF-THE-PRAIRIE.

† **Filipendula ulmaria** (L.) Maxim. QUEEN-OF-THE-MEADOW.

† **Fragaria chiloensis** (L.) Mill. BEACH STRAWBERRY.

Fragaria vesca L. WOODLAND STRAWBERRY, THIN-LEAVED WILD STRAWBERRY.
    var. **americana** Porter (subsp. *americana* (Porter) Staudt)
    * var. **vesca** (subsp. *vesca*)

Fragaria virginiana Duchesne VIRGINIA STRAWBERRY, THICK-LEAVED WILD STRAWBERRY.

Geum aleppicum Jacq. var. **strictum** (Aiton) Fernald YELLOW AVENS.

Geum canadense Jacq. WHITE AVENS.

Geum laciniatum Murray ROUGH AVENS.

Geum rivale L. WATER AVENS, PURPLE AVENS.

Geum vernum (Raf.) Torr. & A. Gray SPRING AVENS.

Geum virginianum L. CREAM-COLORED AVENS.

† **Kerria japonica** (L.) DC. JAPANESE-ROSE.

Malus angustifolia (Aiton) Michx. NARROW-LEAVED CRABAPPLE, SOUTHERN WILD CRABAPPLE. (*Pyrus angustifolia* Aiton)

* **Malus baccata** (L.) Borkh. SIBERIAN CRABAPPLE. (*Pyrus baccata* L.)

Malus coronaria (L.) Mill. WILD CRABAPPLE, SWEET CRABAPPLE. (Incl. var. *dasycalyx* Rehder, var. *lancifolia* (Rehder) C. F. Reed, and M. *glaucescens* Rehder; *Pyrus coronaria* L.)

† **Malus floribunda** Siebold ex Van Houtte SHOWY CRABAPPLE, JAPANESE FLOWERING CRABAPPLE.

Malus ioensis (A. W. Wood) Britton PRAIRIE CRABAPPLE, IOWA CRABAPPLE. (*Pyrus ioensis* (A. W. Wood) L. H. Bailey)

† **Malus pumila** Mill. COMMON APPLE. (*Pyrus malus* L.)

† **Malus sieboldii** (Regel) Rehder TORINGO CRABAPPLE. (*Pyrus sieboldii* Regel)

† **Malus × zumi** (Matsum.) Rehder (**Malus baccata × M. sieboldii**) REDBUD CRABAPPLE, ZUMI CRABAPPLE. (M. *sieboldii* var. *zumi* (Matsum.) Asami)

Malus coronaria × M. ioensis

†**Malus baccata** × **Photinia floribunda**

Photinia floribunda (Lindl.) K. R. Robertson & J. B. Phipps   PURPLE CHOKEBERRY. (*Aronia floribunda* (Lindl.) Spach; *Pyrus floribunda* Lindl.; incl. *Aronia prunifolia* (Marshall) Rehder)

Photinia melanocarpa (Michx.) K. R. Robertson & J. B. Phipps   BLACK CHOKEBERRY. (*Aronia melanocarpa* (Michx.) Elliott; *Pyrus melanocarpa* (Michx.) Willd.)

Physocarpus opulifolius (L.) Maxim.   NINEBARK. (Incl. var. *intermedius* (Rydb.) B. L. Rob.)

Porteranthus stipulatus (Muhl. ex Willd.) Britton   AMERICAN-IPECAC, MIDWEST INDIAN-PHYSIC.

Porteranthus trifoliatus (L.) Britton   BOWMAN'S-ROOT, MOUNTAIN INDIAN-PHYSIC.

Potentilla anserina L.   SILVERWEED. (*Argentina anserina* (L.) Rydb.)

***Potentilla argentea** L.   SILVERY CINQUEFOIL.

Potentilla arguta Pursh   TALL CINQUEFOIL.

Potentilla canadensis L.   RUNNING CINQUEFOIL, DWARF CINQUEFOIL.

Potentilla fruticosa L.   SHRUBBY CINQUEFOIL.

†**Potentilla inclinata** Vill.   GRAY CINQUEFOIL. (Incl. *P. canescens* Besser and Ohio plants treated by some authors as *P. intermedia* L.)

Potentilla norvegica L.   ROUGH CINQUEFOIL, STRAWBERRY-WEED.

Potentilla palustris (L.) Scop.   MARSH CINQUEFOIL, MARSH FIVE-FINGER.

Potentilla paradoxa Nutt.   BUSHY CINQUEFOIL, DIFFUSE CINQUEFOIL.

†**Potentilla pensylvanica** L.   PENNSYLVANIA CINQUEFOIL.

***Potentilla recta** L.   SULPHUR CINQUEFOIL, ROUGH-FRUITED CINQUEFOIL.

†**Potentilla reptans** L.   CREEPING CINQUEFOIL.

Potentilla simplex Michx.   OLD-FIELD CINQUEFOIL, COMMON CINQUEFOIL.

Prunus americana Marshall   WILD PLUM.

***Prunus avium** (L.) L.   SWEET CHERRY.

†**Prunus cerasifera** Ehrh.   CHERRY PLUM, MYROBALAN PLUM.

***Prunus cerasus** L.   SOUR CHERRY, PIE CHERRY.

†**Prunus domestica** L.   COMMON PLUM. (Incl. var. *insititia* (L.) Fiori & Paol.)

Prunus hortulana L. H. Bailey   HORTULAN PLUM, WILD-GOOSE PLUM.

***Prunus mahaleb** L.   MAHALEB CHERRY, PERFUMED CHERRY.

Prunus mexicana S. Watson   BIGTREE PLUM. (*P. americana* var. *lanata* Sudw.)

Prunus munsoniana W. Wight & Hedrick   MUNSON'S WILD-GOOSE PLUM.

Prunus nigra Aiton   CANADA PLUM.

Prunus pensylvanica L.f.   PIN CHERRY, FIRE CHERRY.

†**Prunus persica** (L.) Batsch   PEACH.

Prunus pumila L.
  var. **pumila**   GREAT LAKES SAND CHERRY.
  var. **susquehanae** (Willd.) H. Jaeger   SAND CHERRY. (Incl. *P. pumila* var. *cuneata* (Raf.) L. H. Bailey; *P. susquehanae* Willd.)

Prunus serotina Ehrh.   WILD BLACK CHERRY.

†**Prunus subhirtella** Miq.   HIGAN CHERRY.

†**Prunus tomentosa** Thunb.   DOWNY CHERRY, NANKING CHERRY.

Prunus virginiana L.   CHOKE CHERRY.

†**Pyracantha coccinea** M. Roem.   FIRE-THORN. (*Cotoneaster pyracantha* (L.) Spach)

†**Pyrus calleryana** Decne.   CALLERY PEAR.

†**Pyrus communis** L.   PEAR.

†**Rhodotypos scandens** (Thunb.) Makino   JETBEAD. (*R. tetrapetalus* (Siebold) Makino)

Rosa arkansana Porter **var. suffulta** (Greene) Cockrell   PRAIRIE WILD ROSE, DWARF PRAIRIE ROSE, SUNSHINE ROSE.

Rosa blanda Aiton   SMOOTH ROSE.

***Rosa canina** L.   DOG ROSE.

Rosa carolina L.   PASTURE ROSE, CAROLINA ROSE. (Incl. var. *villosa* (Best) Rehder)

†**Rosa centifolia** L.   CABBAGE ROSE.

†**Rosa cinnamomea** L.   CINNAMON ROSE.

***Rosa eglanteria** L.   EGLANTINE, SWEETBRIER. (Incl. *R. micrantha* Borrer ex. Sm.)

†**Rosa gallica** L.   FRENCH ROSE.

***Rosa multiflora** Thunb. ex Murray   MULTIFLORA ROSE.

†**Rosa nitida** Willd.   NEW ENGLAND ROSE.

Rosa palustris Marshall   SWAMP ROSE.

†**Rosa rugosa** Thunb.   RUGOSA ROSE, JAPANESE ROSE, TURKESTAN ROSE.

Rosa setigera Michx.   CLIMBING PRAIRIE ROSE. (Incl. var. *tomentosa* Torr. & A. Gray)

†**Rosa spinosissima** L.   Scotch Rose, Burnet
   Rose.   (Incl. *R. pimpinellifolia* L.)
†**Rosa wichuraiana** Crép.   Memorial Rose.
Rosa × palustriformis Rydb. (**Rosa blanda ×
   R. palustris**)   Meadow Rose.
Rosa × rudiuscula Greene (**Rosa arkansana ×
   R. carolina**)   Midwest Rose.
Rosa blanda × R. carolina
Rosa carolina × R. palustris
Rubus allegheniensis Porter   Common
   Blackberry.   (Incl. *R. alumnus* L. H. Bailey)
Rubus argutus Link   Southern Blackberry.
Rubus canadensis L.   Smooth Blackberry.
†**Rubus discolor** Weihe & Nees   Himalayan
   Blackberry.   (Incl. *R. procerus* P. J. Müll.)
Rubus enslenii Tratt.   Southern Dewberry.
   (Incl. *R. nefrens* L. H. Bailey, *R. rosagnetis* L. H.
   Bailey, and *R. tetricus* L. H. Bailey)
Rubus flagellaris Willd.   Northern Dewberry.
   (Incl. *R. baileyanus* Britton and *R. roribaccus*
   (L. H. Bailey) Rydb.)
Rubus hispidus L.   Swamp Dewberry.
   *var. hispidus
   var. obovalis (Michx.) Fernald
Rubus idaeus L.   Red Raspberry.
   †var. idaeus (subsp. *idaeus*)
   var. strigosus (Michx.) Maxim. (subsp. *strigo-
      sus* (Michx.) Focke, *R. strigosus* Michx.)
*****Rubus laciniatus** Willd.   Cut-leaved
   Blackberry, Evergreen Blackberry.
Rubus occidentalis L.   Black Raspberry.
Rubus odoratus L.   Flowering Raspberry.
Rubus orarius Blanch.   Northern
   Blackberry.
Rubus pensilvanicus Poir.   Pennsylvania
   Blackberry.   (Incl. *R. bushii* L. H. Bailey,
   R. frondosus Bigelow, and *R. recurvans* Blanch.)
*****Rubus phoenicolasius** Maxim.   Wineberry.
Rubus pubescens Raf.   Dwarf Raspberry.
Rubus setosus Bigelow   Small Bristleberry,
   Bristly Blackberry.
Rubus trivialis Michx.   Southern Dewberry,
   Coastal-plain Dewberry.
Rubus × neglectus Peck (**Rubus idaeus ×
   R. occidentalis**)   Purple Raspberry.
Sanguisorba canadensis L.   Canada Burnet,
   American Burnet.

*****Sanguisorba minor** Scop.   Salad Burnet.
*****Sorbaria sorbifolia** (L.) A. Br.   False Spiraea.
*****Sorbus aucuparia** L.   European Mountain-
   ash, Rowan-tree.   (*Pyrus aucuparia* (L.)
   Gaertn.)
Sorbus decora (Sarg.) C. K. Schneid.   Showy
   Mountain-ash, Western Mountain-ash.
   (*Pyrus decora* (Sarg.) Hyl.)
Spiraea alba Du Roi **var. alba**   Meadow-sweet.
†**Spiraea japonica** L. f.   Japanese Spiraea.
†**Spiraea prunifolia** Siebold & Zucc.   Asian
   Bridal-wreath.
Spiraea tomentosa L.   Hardhack,
   Steeplebush.   (Incl. var. *rosea* (Raf.)
   Fernald)
Spiraea virginiana Britton   Appalachian
   Spiraea.
†**Spiraea × vanhouttei** (Briot) Zabel (**Spiraea
   cantoniensis** Lour. × **S. trilobata** L.)   Bridal-
   wreath.
Waldsteinia fragarioides (Michx.) Tratt.
   Barren-strawberry.

Order **Fabales**

■ MIMOSACEAE.   Mimosa Family

*****Albizia julibrissin** Durazz.   Silk-tree,
   Mimosa-tree.
Desmanthus illinoensis (Michx.) MacMill.
   ex B. L. Rob. & Fernald   Bundleflower,
   Prairie-mimosa.

■ CAESALPINIACEAE.   Caesalpinia Family

Cercis canadensis L.   Redbud.
Chamaecrista fasciculata (Michx.) Greene
   Partridge-pea, Prairie Senna.   (*Cassia
   chamaecrista* L., *Cassia fasciculata* Michx.)
Chamaecrista nictitans (L.) Moench   Wild
   Sensitive-plant.   (*Cassia nictitans* L.)
Gleditsia triacanthos L.   Honey-locust.
Gymnocladus dioica (L.) K. Koch   Kentucky
   Coffee-tree.
Senna hebecarpa (Fernald) Irwin & Barneby
   Northern Wild Senna.   (*Cassia hebecarpa*
   Fernald)

Senna marilandica (L.) Link Southern Wild Senna. (*Cassia marilandica* L.)

■ FABACEAE or PAPILIONACEAE. Pea Family or Bean Family

Amorpha fruticosa L. False Indigo.

Amphicarpaea bracteata (L.) Fernald Hog-peanut. (Incl. var. *comosa* (L.) Fernald and *A. pitcheri* Torr. & A. Gray)

†**Anthyllis vulneraria** L. Kidney-vetch.

Apios americana Medik. Common Groundnut.

Astragalus canadensis L. Canada Milk-vetch.

Astragalus neglectus (Torr. & A. Gray) E. Sheld. Cooper's Milk-vetch.

Baptisia alba (L.) Vent. **var. macrophylla** (Larisey) Isely White False Indigo, Prairie False Indigo. (Incl. *B. lactea* (Raf.) Thieret and *B. leucantha* Torr. & A. Gray)

Baptisia australis (L.) R. Br. Blue False Indigo.

Baptisia tinctoria (L.) R. Br. **var. crebra** Fernald Yellow False Indigo, Horsefly-weed.

†**Cladrastis kentuckea** (Dum. Cours.)Rudd Yellow-wood. (Incl. *C. lutea* (F. Michx.) M. Koch)

Clitoria mariana L. Butterfly-pea.

†**Colutea arborescens** L. Bladder-senna.

*Coronilla varia** L. Crown-vetch.

*Crotalaria sagittalis** L. Rattlebox.

†**Dalea leporina** (Aiton) Bullock Foxtail Dalea, Hare's-foot Dalea. (Incl. *D. alopecuroides* Willd.)

Dalea purpurea Vent. Purple Prairie-clover.

Desmodium canadense (L.) DC. Canada Tick-trefoil.

Desmodium canescens (L.) DC. Hoary Tick-trefoil.

Desmodium ciliare (Muhl. ex Willd.) DC. Little-leaved Tick-trefoil. (Treated by some authors as *D. obtusum* (Muhl. ex Willd.) DC.)

Desmodium cuspidatum (Muhl. ex Willd.) DC. ex Loudon Big Tick-trefoil. (Incl. var. *longifolium* (Torr. & A. Gray) B. G. Schub.)

Desmodium glabellum (Michx.) DC. Hairy Tick-trefoil.

Desmodium glutinosum (Muhl. ex Willd.) A. W. Wood Cluster-leaved Tick-trefoil.

Desmodium illinoense A. Gray Prairie Tick-trefoil.

Desmodium laevigatum (Nutt.) DC. Smooth Tick-trefoil.

Desmodium marilandicum (L.) DC. Maryland Tick-trefoil.

Desmodium nudiflorum (L.) DC. Naked Tick-trefoil.

Desmodium paniculatum (L.) DC. Panicled Tick-trefoil.

Desmodium pauciflorum (Nutt.) DC. Few-flowered Tick-trefoil.

Desmodium perplexum B. G. Schub. Erect Tick-trefoil. (Incl. *D. dillenii* Darl.)

Desmodium rigidum (Elliott) DC. Stiff Tick-trefoil. (Incl. *D. obtusum* (Muhl. ex Willd.) DC.)

Desmodium rotundifolium DC. Round-leaved Tick-trefoil.

Desmodium sessilifolium (Torr.) Torr. & A. Gray Sessile Tick-trefoil, Sessile-leaved Tick-trefoil.

Desmodium viridiflorum (L.) DC. Velvety Tick-trefoil. (Incl. *D. nuttallii* (Schindl.) B. G. Schub.)

†**Dolichos lablab** L. Hyacinth Bean, Lablab Bean. (*Lablab purpureus* (L.) Sweet)

Galactia volubilis (L.) Britton Milk-pea, Hairy Milk-pea.

†**Genista tinctoria** L. Dyer's Greenweed.

†**Glycine max** (L.) Merr. Soybean.

*Kummerowia stipulacea** (Maxim.) Makino Korean-clover. (*Lespedeza stipulacea* Maxim.)

*Kummerowia striata** (Thunb.) Schindl. Japanese-clover. (*Lespedeza striata* (Thunb.) Hook. & Arn.)

Lathyrus japonicus Willd. Inland Beach Pea. (Treated by some authors as a part of *L. maritimus* (L.) Bigelow)

*Lathyrus latifolius** L. Perennial Pea, Everlasting Pea.

Lathyrus ochroleucus Hook.   Yellow
  Vetchling, White Pea.

†Lathyrus odoratus L.   Sweet Pea.

Lathyrus palustris L.   Marsh Pea.

†Lathyrus pratensis L.   Meadow Pea, Yellow
  Vetchling.

*Lathyrus tuberosus L.   Tuberous Vetchling,
  Earth-nut Pea.

Lathyrus venosus Muhl. ex Willd.   Wild Pea,
  Forest Pea.

†Lespedeza bicolor Turcz.   Bicolored
  Lespedeza.

Lespedeza capitata Michx.   Bush-clover,
  Round-headed Bush-clover, Round-
  headed Lespedeza.

*Lespedeza cuneata (Dumont) G. Don
  Chinese Lespedeza.

†Lespedeza formosa (Vogel) Koehne
  Formosan Lespedeza.

Lespedeza hirta (L.) Hornem.   Hairy
  Lespedeza.

Lespedeza intermedia (S. Watson) Britton
  Wand Lespedeza.

Lespedeza procumbens Michx.   Downy
  Trailing Lespedeza.

Lespedeza repens (L.) Barton   Smooth
  Trailing Lespedeza.

†Lespedeza thunbergii (DC.) Nakai   Tall
  Lespedeza.

Lespedeza violacea (L.) Pers.   Violet
  Lespedeza.

Lespedeza virginica (L.) Britton   Virginia
  Lespedeza, Slender Lespedeza.

Lespedeza × nuttallii Darl. (Lespedeza hirta ×
  L. intermedia)   Nuttall's Lespedeza.

Lespedeza hirta × L. virginica

Lespedeza repens × L. virginica

*Lotus corniculatus L.   Birdfoot-trefoil.

Lupinus perennis L.   Wild Lupine, Sundial
  Lupine.

*Medicago lupulina L.   Black Medick.

*Medicago sativa L.
  *subsp. falcata (L.) Arcang.   Yellow
    Alfalfa.   (M. falcata L.)
  *subsp. sativa   Alfalfa.

*Melilotus albus Medik.   White Sweet-
  clover.

*Melilotus altissimus Thuill.   Tall Sweet-
  clover.

*Melilotus officinalis (L.) Pall.   Yellow Sweet-
  clover.

Orbexilum onobrychis (Nutt.) Rydb.   Sainfoin
  Scurf-pea.   (Psoralea onobrychis Nutt.)

Orbexilum pedunculatum (Mill.) Rydb.   False
  Scurf-pea.   (Incl. Psoralea psoralioides
  (Walter) Cory)

†Phaseolus coccineus L.   Scarlet Runner
  Bean.

Phaseolus polystachios (L.) Britton, Sterns &
  Poggenb.   Wild Bean, Wild Kidney Bean.

†Phaseolus vulgaris L.   Green Bean, Bush
  Bean.

*Pueraria lobata (Willd.) Ohwi   Kudzu,
  Kudzu-vine.   (P. montana (Lour.) Merr.
  var. lobata (Willd.) Maesen & S. M. Almeida)

*Robinia hispida L.   Bristly Locust, Rose-
  acacia.

Robinia pseudoacacia L.   Black Locust.

*Robinia viscosa Vent.   Clammy Locust.

*Robinia × ambigua Poir. (Robinia hispida ×
  R. pseudoacacia)   Pink Locust.

Strophostyles helvola (L.) Elliott   Annual
  Woolly-bean.

†Strophostyles leiosperma (Torr. & A. Gray)
  Piper   Small-flowered Woolly-bean.

Strophostyles umbellata (Muhl. ex Willd.)
  Britton   Perennial Woolly-bean.

Stylosanthes biflora (L.) Britton, Sterns &
  Poggenb.   Pencil-flower.   (Incl. S. riparia
  Kearney)

†Styphnolobium japonicum (L.) Schott
  Japanese Pagoda-tree, Chinese Scholar-
  tree.   (Sophora japonica L.)

Tephrosia virginiana (L.) Pers.   Goat's-rue,
  Rabbit-pea.

*Trifolium arvense L.   Rabbit's-foot Clover.

*Trifolium aureum Pollich   Yellow Clover,
  Palmate Hop Clover.   (Treated by some
  authors as a part of T. agrarium L.)

*Trifolium campestre Schreb.   Low Hop
  Clover, Pinnate Hop Clover.   (Treated
  by some authors as a part of T. procumbens L.)

*Trifolium dubium Sibth.   Little Hop Clover.

*Trifolium hybridum L.   Alsike Clover.

†**Trifolium incarnatum** L.   Crimson
    Clover.
***Trifolium pratense** L.   Red Clover.
Trifolium reflexum L.   Buffalo Clover,
    Annual Buffalo Clover.   (Incl. var.
    glabrum Lojac.)
***Trifolium repens** L.   White Clover.
†**Trifolium resupinatum** L.   Persian Clover.
Trifolium stoloniferum Muhl. ex Eaton
    Running Buffalo Clover.
Vicia americana Muhl. ex Willd.   American
    Vetch.
Vicia caroliniana Walter   Pale Vetch, Wood
    Vetch.
***Vicia cracca** L.   Bird Vetch, Tufted Vetch.
***Vicia dasycarpa** Ten.   Woolly-podded Vetch.
    (*V. villosa* var. *glabrescens* W. D. J. Koch, *V. vil-
    losa* subsp. *varia* (Host) Corb.)
***Vicia hirsuta** (L.) Gray   Tiny Vetch.
***Vicia sativa** L.   Spring Vetch, Common
    Vetch.
    ***var. nigra** L. (subsp. *nigra* (L.) Ehrh.; incl.
        V. angustifolia L.)
    †**var. sativa** (subsp. *sativa*)
***Vicia tetrasperma** (L.) Schreb.   Sparrow
    Vetch, Four-seeded Vetch.
***Vicia villosa** Roth   Hairy Vetch.
†**Vigna unguiculata** (L.) Walp.   Cow-pea,
    Black-eyed-pea.
†**Wisteria floribunda** (Willd.) DC.   Japanese
    Wisteria.
Wisteria frutescens (L.) Poir.   Atlantic
    Wisteria.   (Incl. *W. macrostachya* (Torr. &
    A. Gray) Nutt. ex B. L. Rob. & Fernald)

Order **Proteales**

■ ELAEAGNACEAE.   Oleaster Family

***Elaeagnus angustifolia** L.   Russian-olive,
    Oleaster.
†**Elaeagnus multiflora** Thunb.   Long-stalked
    Oleaster.
***Elaeagnus umbellata** Thunb.   Autumn-
    olive.
Shepherdia canadensis (L.) Nutt.   Buffalo-
    berry, Canada Buffalo-berry.

Order **Podostemales**

■ PODOSTEMACEAE.   Riverweed Family

Podostemum ceratophyllum Michx.
    Riverweed.

Order **Haloragales**

■ HALORAGACEAE.   Water-milfoil
    Family

†**Myriophyllum aquaticum** (Vell.) Verdc.
    Parrot's-feather.
Myriophyllum heterophyllum Michx.   Broad-
    leaved Water-milfoil, Two-leaved Water-
    milfoil.
†**Myriophyllum humile** (Raf.) Morong   Lowly
    Water-milfoil.
†**Myriophyllum pinnatum** (Walter) Britton,
    Sterns & Poggenb.   Pinnate Water-milfoil.
Myriophyllum sibiricum Kom.   Northern
    Water-milfoil, American Water-milfoil.
    (Incl. *M. exalbescens* Fernald; *M. spicatum* var.
    squamosum (Laest. ex Hartm.) Hartm.)
***Myriophyllum spicatum** L.   Eurasian Water-
    milfoil, Spiked Water-milfoil.
Myriophyllum verticillatum L.   Whorled
    Water-milfoil, Green Water-milfoil.
Proserpinaca palustris L. **var. crebra** Fernald &
    Griscom   Mermaid-weed.

Order **Myrtales**

■ LYTHRACEAE.   Loosestrife Family

Ammannia coccinea Rottb.   Long-leaved
    Ammannia.
Ammannia robusta Heer & Regel   Sessile-
    fruited Ammannia.
Cuphea viscosissima Jacq.   Clammy Cuphea,
    Blue Waxweed.
Decodon verticillatus (L.) Elliott   Swamp
    Loosestrife, Water-willow.   (Incl. var.
    laevigatus Torr. & A. Gray)
Lythrum alatum Pursh   Wing-angled
    Loosestrife.
†**Lythrum hyssopifolia** L.   Hyssop-leaved
    Loosestrife.

*Lythrum salicaria L.   Purple Loosestrife, Spiked Loosestrife.
Rotala ramosior (L.) Koehne   Tooth-cup.

■ THYMELAEACEAE.   Mezereum Family

†Daphne mezereum L.   Mezereon.
Dirca palustris L.   Leatherwood.
*Thymelaea passerina (L.) Coss. & Germ. Thymelaea.

■ ONAGRACEAE.   Evening-primrose Family

Circaea alpina L.   Small Enchanter's-nightshade.
Circaea lutetiana L. var. canadensis L. Common Enchanter's-nightshade. (subsp. canadensis (L.) Asch. & Magnus)
Circaea × intermedia Ehrh. (Circaea alpina × C. lutetiana)   Intermediate Enchanter's-nightshade.
†Clarkia pulchella Pursh   Farewell-to-spring.
Epilobium angustifolium L. var. platyphyllum (Daniels) Fernald   Fireweed.   (subsp. circumvagum Mosquin; incl. var. canescens A. W. Wood)
Epilobium ciliatum Raf.   Northern Willow-herb, Simple Willow-herb.
Epilobium coloratum Biehler   Purple-leaved Willow-herb.
*Epilobium hirsutum L.   Hairy Willow-herb.
Epilobium leptophyllum Raf.   Linear-leaved Willow-herb.
*Epilobium parviflorum Schreb.   Small-flowered Hairy Willow-herb.
Epilobium strictum Muhl. ex Spreng.   Downy Willow-herb, Simple Willow-herb.
Epilobium × wisconsinense Ugent (Epilobium ciliatum × E. coloratum)   Wisconsin Willow-herb.
Gaura biennis L.   Biennial Gaura.
†Gaura longiflora Spach   Western Gaura.   (G. biennis var. pitcheri Torr. & A. Gray)
†Gaura parviflora Douglas ex Lehm.   Small-flowered Gaura.
Ludwigia alternifolia L.   Seedbox.

*Ludwigia decurrens Walter   Erect Primrose-willow.
*Ludwigia leptocarpa (Nutt.) Hara   Hairy Primrose-willow.
Ludwigia palustris (L.) Elliott   Water-purslane, Marsh-purslane.   (Incl. var. americana (DC.) Fernald & Griscom)
*Ludwigia peploides (Kunth) P. H. Raven var. glabrescens (Kuntze) Shinners   Creeping Primrose-willow.   (subsp. glabrescens (Kuntze) P. H. Raven)
Ludwigia polycarpa Short & R. Peter   False Loosestrife.
Oenothera biennis L.
   var. biennis   Common Evening-primrose.
   var. canescens Torr. & A. Gray   Hoary Evening-primrose.   (O. villosa Thunb.)
   var. nutans (G. F. Atk. & Bartlett) Wiegand   Appalachian Evening-primrose.   (Incl. var. austromontana (Munz) Cronquist; O. nutans G. F. Atk. & Bartlett)
Oenothera clelandii W. Dietr., P. H. Raven & W. L. Wagner   Cleland's Evening-primrose.
Oenothera fruticosa L.   Common Sundrops.
   var. ambigua Nutt. (subsp. glauca (Michx.) Straley; O. tetragona Roth)
   var. fruticosa (subsp. fruticosa)
Oenothera laciniata Hill   Cut-leaved Evening-primrose.
Oenothera oakesiana (A. Gray) J. W. Robbins ex S. Watson & J. M. Coult.   Oakes' Evening-primrose.   (O. biennis var. oakesiana A. Gray; O. parviflora var. oakesiana (A. Gray) Fernald)
Oenothera parviflora L.   Small-flowered Evening-primrose.   (O. biennis var. parviflora (L.) Torr. & A. Gray)
Oenothera perennis L.   Small Sundrops.
Oenothera pilosella Raf.   Meadow Sundrops.
†Oenothera speciosa Nutt.   White Evening-primrose.
†Oenothera triloba Nutt.   Sessile Evening-primrose.
†Stenosiphon linifolius (Nutt. ex James) Heynh.   False Gaura.

■ MELASTOMATACEAE. Melastome Family

Rhexia virginica L. Virginia Meadow-beauty.

Order **Cornales**

■ CORNACEAE. Dogwood Family

Cornus alternifolia L. f. Alternate-leaved Dogwood, Pagoda Dogwood.
Cornus amomum Mill. Silky Dogwood.
  var. amomum (subsp. *amomum*)
  var. schuetzeana (C. A. Mey.) Rickett (subsp.
    obliqua (Raf.) J. S. Wilson, *C. obliqua* Raf.)
Cornus canadensis L. Bunchberry, Dwarf Cornel.
Cornus drummondii C. A. Mey. Rough-leaved Dogwood.
Cornus florida L. Flowering Dogwood.
Cornus racemosa Lam. Gray Dogwood, Panicled Dogwood.
Cornus rugosa Lam. Round-leaved Dog-wood.
Cornus stolonifera Michx. Red Osier, Red Osier Dogwood. (Incl. var. *baileyi* (J. M. Coult. & W. H. Evans) Drescher; treated by some authors as a part of *C. sericea* L.)
Cornus × arnoldiana Rehder (**Cornus amomum var. schuetzeana × C. racemosa**) Arnold Arboretum Dogwood.
Nyssa sylvatica Marshall Sour-gum, Black-gum, Tupelo. (Incl. var. *caroliniana* (Poir.) Fernald) Often placed in a segregate family, the *Nyssaceae*.

Order **Santalales**

■ SANTALACEAE. Sandalwood Family

Comandra umbellata (L.) Nutt. Bastard-toadflax, Star-toadflax.

■ VISCACEAE. Christmas-mistletoe Family

Phoradendron leucarpum (Raf.) Reveal & M. C. Johnst. American Mistletoe, American Christmas-mistletoe. (Incl. *P. flavescens* Nutt. ex Engelm. and *P. serotinum* (Raf.) M. C. Johnst.)

Order **Celastrales**

■ CELASTRACEAE. Bittersweet Family or Staff-tree Family

*Celastrus orbiculatus Thunb. Oriental Bittersweet, Asian Bittersweet, Chinese Bittersweet.
Celastrus scandens L. Bittersweet, American Bittersweet, Waxwork.
*Euonymus alatus (Thunb.) Siebold Winged Wahoo, Winged Spindle-tree.
Euonymus americanus L. American Strawberry-bush, Hearts-bursting-open-with-love.
Euonymus atropurpureus Jacq. Wahoo, Burning-bush.
*Euonymus europaeus L. European Spindle-tree.
*Euonymus fortunei (Turcz.) Hand.-Mazz. Wintercreeper, Chinese Wintercreeper.
Euonymus obovatus Nutt. Running Strawberry-bush.
Paxistima canbyi A. Gray Cliff-green, Canby's Mountain-lover.

■ AQUIFOLIACEAE. Holly Family

Ilex opaca Aiton American Holly.
Ilex verticillata (L.) A. Gray Winterberry, Winterberry Holly, Michigan Holly. (Incl. var. *padifolia* (Willd.) Torr. & A. Gray ex S. Watson and var. *tenuifolia* (Torr.) S. Watson)
Nemopanthus mucronatus (L.) Loes. Mountain-holly, Catberry.

Order **Euphorbiales**

■ BUXACEAE. Boxwood Family

†Buxus sempervirens L. Common Boxwood.
†Pachysandra terminalis Siebold & Zucc. Japanese Pachysandra.

- EUPHORBIACEAE. SPURGE FAMILY

†**Acalypha gracilens** A. Gray SLENDER THREE-SEEDED MERCURY.

***Acalypha ostryifolia** Riddell HORNBEAM THREE-SEEDED MERCURY.

Acalypha virginica L.
    var. deamii Weath. DEAM'S THREE-SEEDED MERCURY. (*A. deamii* (Weath.) Ahles)
    var. rhomboidea (Raf.) Cooperr. RHOMBIC THREE-SEEDED MERCURY. (*A. rhomboidea* Raf.)
    var. virginica VIRGINIA THREE-SEEDED MERCURY.

***Croton capitatus** Michx. WOOLLY CROTON, HOGWORT.

Croton glandulosus L. **var. septentrionalis** Müll. Arg. GLANDULAR CROTON, SAND CROTON, NORTHERN CROTON.

***Croton monanthogynus** Michx. PRAIRIE-TEA.

Euphorbia commutata Engelm. WOOD SPURGE.

Euphorbia corollata L. FLOWERING SPURGE.

†**Euphorbia cyathophora** Murray FIRE-ON-THE-MOUNTAIN, PAINTED-LEAF.

***Euphorbia cyparissias** L. CYPRESS SPURGE.

***Euphorbia dentata** Michx. TOOTHED SPURGE, WILD POINSETTIA.

***Euphorbia esula** L. LEAFY SPURGE.

***Euphorbia falcata** L. FALCATE SPURGE.

†**Euphorbia helioscopia** L. SUN SPURGE.

Euphorbia humistrata Engelm. SPREADING SPURGE. (*Chamaesyce humistrata* (Engelm.) Small)

***Euphorbia lathyris** L. CAPER SPURGE.

Euphorbia maculata L. PROSTRATE SPURGE. (Incl. *E. supina* Raf.; *Chamaesyce maculata* (L.) Small)

***Eurphorbia marginata** Pursh SNOW-ON-THE-MOUNTAIN.

Euphorbia nutans Lag. SPOTTED SPURGE, NODDING SPURGE. (*Chamaesyce nutans* (Lag.) Small)

Euphorbia obtusata Pursh BLUNT-LEAVED SPURGE.

***Euphorbia peplus** L. PETTY SPURGE.

***Euphorbia platyphyllos** L. BROAD-LEAVED SPURGE.

Euphorbia polygonifolia L. SEASIDE SPURGE. (*Chamaesyce polygonifolia* (L.) Small)

***Euphorbia prostrata** Aiton GROUND-FIG SPURGE. (*Chamaesyce prostrata* (Aiton) Small)

Euphorbia purpurea (Raf.) Fernald GLADE SPURGE.

Euphorbia serpens Kunth ROUND-LEAVED SPURGE. (*Chamaesyce serpens* (Kunth) Small)

Euphorbia vermiculata Raf. HAIRY SPURGE. (*Chamaesyce vermiculata* (Raf.) House)

†**Mercurialis annua** L. HERB MERCURY, BOYS-AND-GIRLS.

Phyllanthus caroliniensis Walter CAROLINA LEAF-FLOWER.

†**Ricinus communis** L. CASTOR-BEAN.

Order **Rhamnales**

- RHAMNACEAE. BUCKTHORN FAMILY

Ceanothus americanus L. NEW JERSEY-TEA. (Incl. var. *pitcheri* Torr. & A. Gray)

Ceanothus herbaceus Raf. NARROW-LEAVED NEW JERSEY-TEA, PRAIRIE REDROOT. (Incl. *C. ovatus* of authors, not Desf.)

Rhamnus alnifolia L'Hér. ALDER-LEAVED BUCKTHORN.

Rhamnus caroliniana Walter CAROLINA BUCKTHORN, INDIAN-CHERRY. (Incl. var. *mollis* Fernald; *Frangula caroliniana* (Walter) A. Gray)

***Rhamnus cathartica** L. COMMON BUCKTHORN, EUROPEAN BUCKTHORN.

†**Rhamnus citrifolia** (Weston) W. J. Hess & Stearn CITRUS-LEAVED BUCKTHORN. (Treated by some authors as a part of *R. davurica* Pall.)

***Rhamnus frangula** L. (*Frangula alnus* Mill.) GLOSSY BUCKTHORN, EUROPEAN ALDER BUCKTHORN, ALDER BUCKTHORN.
    †**forma asplenifolia** (Dippel) Beissn. FERN-LEAVED BUCKTHORN.

Rhamnus lanceolata Pursh LANCE-LEAVED BUCKTHORN. (Incl. var. *glabrata* Gleason)

†**Rhamnus utilis** Decne. CHINESE BUCKTHORN.

■ VITACEAE.   Grape Family

*Ampelopsis brevipedunculata (Maxim.) Trautv.
   Porcelain-berry.
Ampelopsis cordata Michx.   Raccoon-grape.
Parthenocissus inserta (A. Kern.) Fritsch
   Thicket Creeper.   (Incl. *P. vitacea* (Knerr)
   Hitchc.)
Parthenocissus quinquefolia (L.) Planch.
   Virginia Creeper.
†Parthenocissus tricuspidata (Siebold & Zucc.)
   Planch.   Boston Ivy.
Vitis aestivalis Michx.   Summer Grape.
   var. aestivalis
   var. bicolor Deam (Incl. var. *argentifolia*
      (Munson) Fernald)
Vitis cinerea (Engelm.) Millard var. baileyana
   (Munson) Comeaux   Pigeon Grape.
   (*V. baileyana* Munson)
Vitis labrusca L.   Fox Grape, Northern Fox
   Grape.   (Incl. *V. labruscana* L. H. Bailey)
Vitis riparia Michx.   Riverbank Grape.
Vitis vulpina L.   Frost Grape.
Vitis aestivalis × V. cinerea

Order **Linales**

■ LINACEAE.   Flax Family

†Linum grandiflorum Desf.   Flowering Flax.
Linum medium (Planch.) Britton var. texanum
   (Planch.) Fernald   Stiff Yellow Flax.
†Linum perenne L.   Perennial Flax.   (Incl.
   L. lewisii Pursh)
Linum striatum Walter   Ridged Yellow Flax.
Linum sulcatum Riddell   Grooved Flax,
   Grooved Yellow Flax.
†Linum usitatissimum L.   Common Flax.
Linum virginianum L.   Slender Yellow Flax.

Order **Polygalales**

■ POLYGALACEAE.   Milkwort Family

Polygala cruciata L. var. aquilonia Fernald &
   B. G. Schub.   Cross-leaved Milkwort.
Polygala curtissii A. Gray   Curtiss' Milkwort.
Polygala incarnata L.   Pink Milkwort.

Polygala paucifolia Willd.   Fringed Milkwort,
   Gay-wings.
Polygala polygama Walter   Racemed
   Milkwort.   (Incl. var. *obtusata* Chodat)
Polygala sanguinea L.   Purple Milkwort, Field
   Milkwort.
Polygala senega L.   Seneca Snakeroot.   (Incl.
   var. *latifolia* Torr. & A. Gray)
Polygala verticillata L.   Whorled Milkwort.
   var. ambigua (Nutt.) A. W. Wood (*P. ambigua*
      Nutt.)
   var. verticillata (Incl. var. *isocycla* Fernald and
      var. *sphenostachya* Pennell)

Order **Sapindales**

■ STAPHYLEACEAE.   Bladdernut Family

Staphylea trifolia L.   Bladdernut.

■ SAPINDACEAE.   Soapberry Family

†Cardiospermum halicacabum L.   Balloon-
   vine.
†Koelreuteria paniculata Laxm.   Golden-
   raintree.

■ HIPPOCASTANACEAE.   Horse-
chestnut Family

Aesculus flava Aiton   Sweet Buckeye, Yellow
   Buckeye.   (Incl. A. *octandra* Marshall)
Aesculus glabra Willd.   Ohio Buckeye.   (Incl.
   var. *sargentii* Rehder)
†Aesculus hippocastanum L.   Horse-
   chestnut.
Aesculus × marylandica Booth ex Kirchn.
   (Aesculus flava × A. glabra)   Maryland
   Buckeye.

■ ACERACEAE.   Maple Family

†Acer campestre L.   Hedge Maple.
†Acer ginnala Maxim.   Amur Maple.
Acer negundo L.   Box-elder, Ash-leaved
   Maple.   (Incl. var. *violaceum* (Kirchn.)
   Jaeger)

†**Acer palmatum** Thunb.   Japanese Maple.
Acer pensylvanicum L.   Striped Maple,
   Moosewood.
*****Acer platanoides** L.   Norway Maple.
Acer rubrum L.   Red Maple.
   var. rubrum
   var. trilobum Torr. & A. Gray ex K. Koch
Acer saccharinum L.   Silver Maple.
Acer saccharum Marshall
   var. saccharum   Sugar Maple.   (subsp.
   saccharum)
   var. viride (Schmidt) E. Murray   Black
   Maple.   (subsp. *nigrum* (F. Michx.)
   Desmarais, A. *nigrum* F. Michx.)
Acer spicatum Lam.   Mountain Maple.
†**Acer tataricum** L.   Tatarian Maple.
Acer × freemanii E. Murray (**Acer rubrum ×
   A. saccharinum**)   Freeman's Maple.

■   ANACARDIACEAE.   Sumac Family or
   Cashew Family

†**Cotinus coggygria** Scop.   Smoke-tree.
Rhus aromatica Aiton
   var. arenaria (Greene) Fernald   Beach
   Sumac.
   var. aromatica   Fragrant Sumac.
Rhus copallina L. **var. latifolia** Engl.   Shining
   Sumac, Winged Sumac.
Rhus glabra L.   Smooth Sumac.
Rhus typhina L.   Staghorn Sumac.   (*R. hirta*
   (L.) Sudw.)
   †**forma dissecta** Rehder   Cut-leaved
   Staghorn Sumac.
Rhus × pulvinata Greene (**Rhus glabra × R. ty-
   phina**)   Greene's Sumac.   (Incl. *R. × bore-
   alis* (Britton) Greene, *R. glabra* var. *borealis*
   Britton)
Toxicodendron radicans (L.) Kuntze (*Rhus radi-
   cans* L.)
   var. radicans   Poison Ivy.   (Incl. var.
   negundo (Greene) Reveal)
   var. rydbergii (Small ex Rydb.) Erskine
   Northern Poison Ivy.   (*T. rydbergii*
   (Small ex Rydb.) Greene)
Toxicodendron vernix (L.) Kuntze   Poison
   Sumac.   (*Rhus vernix* L.)

■   SIMAROUBACEAE.   Quassia Family
*****Ailanthus altissima** (Mill.) Swingle   Tree-of-
   heaven.

■   RUTACEAE.   Rue Family
Ptelea trifoliata L.   Hop-tree, Wafer-ash.
Zanthoxylum americanum Mill.   Prickly-
   ash.

■   ZYGOPHYLLACEAE.   Creosote-bush
   Family
*****Tribulus terrestris** L.   Puncture-vine,
   Puncture-weed.

Order **Geraniales**

■   OXALIDACEAE.   Wood-sorrel Family
Oxalis acetosella L.   White Wood-sorrel.
   (Incl. *O. montana* Raf.)
*****Oxalis corniculata** L.   Creeping Wood-
   sorrel, Creeping Lady's-sorrel.
Oxalis dillenii Jacq.   Southern Yellow Wood-
   sorrel.
Oxalis grandis Small   Great Yellow Wood-
   sorrel.
Oxalis stricta L.   Common Yellow Wood-
   sorrel.
Oxalis violacea L.   Violet Wood-sorrel.

■   GERANIACEAE.   Geranium Family
*****Erodium cicutarium** (L.) L'Hér. ex Aiton
   Alfilaria, Red-stemmed Filaree.
Geranium bicknellii Britton   Bicknell's
   Crane's-bill.
Geranium carolinianum L.   Carolina
   Crane's-bill.   (Incl. var. *confertiflorum*
   Fernald)
*****Geranium columbinum** L.   Long-stalked
   Crane's-bill.
†**Geranium dissectum** L.   Cut-leaved
   Crane's-bill.
Geranium maculatum L.   Wild Geranium,
   Wild Crane's-bill.

*Geranium molle L. Dove's-foot Crane's-bill. (Incl. var. *aequale* Bab.)

*Geranium pusillum L. Small-flowered Crane's-bill.

Geranium robertianum L. Herb Robert.

*Geranium sanguineum L. Blood-red Crane's-bill.

■ LIMNANTHACEAE. Meadow-foam Family

Floerkea proserpinacoides Willd. False Mermaid.

■ BALSAMINACEAE. Touch-me-not Family

† Impatiens balsamina L. Garden Balsalm.

Impatiens capensis Meerb. Spotted Touch-me-not, Orange Jewelweed.

Impatiens pallida Nutt. Pale Touch-me-not, Yellow Jewelweed.

Order **Araliales**

■ ARALIACEAE. Ginseng Family

*Aralia elata (Miq.) Seem. Japanese-angelica.

Aralia hispida Vent. Bristly Sarsaparilla.

Aralia nudicaulis L. Wild Sarsaparilla.

Aralia racemosa L. Spikenard.

Aralia spinosa L. Devil's Walkingstick, Hercules' Club.

*Hedera helix L. English Ivy.

Panax quinquefolius L. American Ginseng.

Panax trifolius L. Dwarf Ginseng.

■ APIACEAE or UMBELLIFERAE. Carrot Family

*Aegopodium podagraria L. Goutweed.

*Aethusa cynapium L. Fool's-parsley.

† Anethum graveolens L. Dill.

Angelica atropurpurea L. Purple-stemmed Angelica.

Angelica venenosa (Greenway) Fernald Hairy Angelica.

*Anthriscus caucalis M. Bieb. Bur Chervil.

*Anthriscus sylvestris (L.) Hoffm. Wild Chervil.

† Apium graveolens L. Celery.

† Bupleurum lancifolium Hornem. Mediterranean Thoroughwax.

† Bupleurum rotundifolium L. Thoroughwax.

† Carum carvi L. Caraway.

† Centella asiatica (L.) Urb. Spadeleaf.

Chaerophyllum procumbens (L.) Crantz
var. procumbens Spreading Chervil.
var. shortii Torr. & A. Gray Short's Chervil.

Cicuta bulbifera L. Bulblet-bearing Water-hemlock.

Cicuta maculata L. Water-hemlock, Spotted Water-hemlock.

Conioselinum chinense (L.) Britton, Sterns & Poggenb. Hemlock-parsley.

*Conium maculatum L. Poison-hemlock.

† Coriandrum sativum L. Coriander.

Cryptotaenia canadensis (L.) DC. Honewort.

*Daucus carota L. Wild Carrot, Queen Anne's Lace.

Erigenia bulbosa (Michx.) Nutt. Harbinger-of-spring, Pepper-and-salt.

Eryngium yuccifolium Michx. Rattlesnake-master.

† Foeniculum vulgare Mill. Fennel.

Heracleum maximum W. Bartram
Cow-parsnip, Masterwort. (Incl. *H. lanatum* Michx.; *H. sphondylium* L. subsp. *montanum* (Schleich. ex Gaudin) Briq.)

Hydrocotyle americana L. American Water-pennywort.

*Hydrocotyle ranunculoides L. f. Floating Water-pennywort.

*Hydrocotyle sibthorpioides Lam. Lawn Water-pennywort.

Hydrocotyle umbellata L. Umbellate Water-pennywort, Navelwort.

† Levisticum officinale W. D. J. Koch European Lovage.

Ligusticum canadense (L.) Britton American Lovage, Nondo.

† Oenanthe aquatica (L.) Poir. Water-fennel.

Osmorhiza claytonii (Michx.) C. B. Clarke Woolly Sweet Cicely.

Osmorhiza longistylis (Torr.) DC.   Smooth
  Sweet Cicely.
Oxypolis rigidior (L.) Raf.   Cowbane.
*Pastinaca sativa L.   Wild Parsnip.
Perideridia americana (Nutt. ex DC.) Rchb.
  Perideridia.
† Petroselinum crispum (Mill.) Nyman ex A. W.
  Hill   Parsley.
Sanicula canadensis L.   Short-styled
  Snakeroot.   (Incl. var. *grandis* Fernald)
Sanicula gregaria E. P. Bicknell   Clustered
  Snakeroot.   (Treated by some authors as
  S. odorata (Raf.) K. M. Pryer & L. R. Phillippe)
Sanicula marilandica L.   Black Snakeroot.
Sanicula trifoliata E. P. Bicknell   Large-
  fruited Snakeroot.
† Scandix pecten-veneris L.   Venus'-comb,
  Shepherd's-needle.
Sium suave Walter   Water-parsnip.
Taenidia integerrima (L.) Drude   Yellow-
  pimpernel.
Thaspium barbinode (Michx.) Nutt.   Hairy-
  jointed Meadow-parsnip.   (Incl. var. *angus-
  tifolium* J. M. Coult. & Rose)
Thaspium trifoliatum (L.) A. Gray
  var. flavum S. F. Blake   Yellow Meadow-
    parsnip.   (Treated by some authors as
    T. trifoliatum var. *aureum* (L.) Britton)
  var. trifoliatum   Purple Meadow-parsnip.
*Torilis arvensis (Huds.) Link   Field Hedge-
  parsley.
*Torilis japonica (Houtt.) DC.   Japanese
  Hedge-parsley.
Zizia aptera (A. Gray) Fernald   Heart-leaved
  Meadow-parsnip.
Zizia aurea (L.) W. D. J. Koch   Golden
  Alexanders, Early Meadow-parsnip.

Subclass **ASTERIDAE**

Order **Gentianales**

■ GENTIANACEAE.   Gentian Family

Bartonia virginica (L.) Britton, Sterns &
  Poggenb.   Yellow Bartonia.
*Centaurium erythraea Rafn   European
  Centaury, Forking Centaury.

*Centaurium pulchellum (Sw.) Druce
  Branching Centaury.
Gentiana alba Muhl. ex Nutt.   Yellowish
  Gentian.   (Incl. G. *flavida* A. Gray)
Gentiana andrewsii Griseb.   Bottle Gentian,
  Closed Gentian.
Gentiana clausa Raf.   Blind Gentian, Closed
  Gentian.
Gentiana puberulenta J. S. Pringle   Prairie
  Gentian, Downy Gentian.
Gentiana saponaria L.   Soapwort Gentian.
Gentiana villosa L.   Sampson's Snakeroot.
Gentiana × billingtonii Farw. (**Gentiana
  andrewsii** × **G. puberulenta**)   Billington's
  Gentian.
Gentianella quinquefolia (L.) Small var. occi-
  dentalis (A. Gray) Small   Stiff Gentian,
  Agueweed.   (subsp. *occidentalis* (A. Gray)
  J. M. Gillett; *Gentiana quinquefolia* L.)
Gentianopsis crinita (Froel.) Ma   Fringed
  Gentian, Eastern Fringed Gentian.
  (*Gentiana crinita* Froel.)
Gentianopsis procera (T. Holm) Ma   Small
  Fringed Gentian, Western Fringed
  Gentian.   (*Gentiana procera* T. Holm)
Obolaria virginica L.   Pennywort.
Sabatia angularis (L.) Pursh   Rose-pink,
  Marsh-pink.
Swertia caroliniensis (Walter) Kuntze
  American Columbo.   (*Frasera caroliniensis*
  Walter)

■ APOCYNACEAE.   Dogbane Family

† Amsonia tabernaemontana Walter var. salici-
  folia (Pursh) Woodson   Blue-star.
Apocynum androsaemifolium L.   Spreading
  Dogbane.
Apocynum cannabinum L.   Indian-hemp.
Apocynum sibiricum Jacq.   Clasping-leaved
  Dogbane.   (Incl. var. *cordigerum* (Greene)
  Fernald; *A. cannabinum* var. *hypericifolium*
  A. Gray)
Apocynum × floribundum Greene (**Apocynum
  androsaemifolium** × **A. cannabinum**)
  Intermediate Dogbane.   (Incl. A. × *medium*
  Greene)

*Vinca major L.   LARGE PERWINKLE, GREATER PERIWINKLE.
*Vinca minor L.   COMMON PERIWINKLE, MYRTLE.

■ ASCLEPIADACEAE.   MILKWEED FAMILY

Asclepias amplexicaulis Sm.   CLASPING-LEAVED MILKWEED, BLUNT-LEAVED MILKWEED.
Asclepias exaltata L.   POKE MILKWEED.
Asclepias hirtella (Pennell) Woodson   GREEN MILKWEED.
Asclepias incarnata L.   SWAMP MILKWEED, PINK MILKWEED.
Asclepias purpurascens L.   PURPLE MILKWEED.
Asclepias quadrifolia Jacq.   FOUR-LEAVED MILKWEED.
Asclepias sullivantii Engelm. ex A. Gray   PRAIRIE MILKWEED, SULLIVANT'S MILKWEED.
Asclepias syriaca L.   COMMON MILKWEED.
Asclepias tuberosa L.   BUTTERFLY-WEED, BUTTERFLY MILKWEED, PLEURISY-ROOT. (Incl. var. *interior* (Woodson) Shinners)
Asclepias variegata L.   WHITE MILKWEED.
Asclepias verticillata L.   WHORLED MILKWEED, HORSETAIL MILKWEED.
Asclepias viridiflora Raf.   GREEN-FLOWERED MILKWEED, GREEN MILKWEED.
Asclepias viridis Walter   SPIDER MILKWEED, ANTELOPE-HORN.
Cynanchum laeve (Michx.) Pers.   SAND-VINE, HONEY-VINE.   (*Ampelamus albidus* (Nutt.) Britton)
Matelea obliqua (Jacq.) Woodson   ANGLE-POD.
*Vincetoxicum nigrum (L.) Moench   BLACK SWALLOW-WORT.   (*Cynanchum louiseae* Kartesz & Gandhi)

Order **Solanales**

■ SOLANACEAE.   NIGHTSHADE FAMILY

†Browallia americana L.   BUSH-VIOLET.
†Datura inoxia Mill.   DOWNY THORN-APPLE, ANGEL'S-TRUMPET.
*Datura stramonium L.   JIMSONWEED.

*Lycium barbarum L.   COMMON MATRIMONY-VINE.   (Incl. *L. chinense* Mill. and *L. halimifolium* Mill.)
†Lycopersicon esculentum Mill.   TOMATO.
*Nicandra physalodes (L.) Gaertn.   APPLE-OF-PERU.
†Nicotiana alata Link & Otto   FLOWERING TO-BACCO.
†Nicotiana glauca Graham   TREE TOBACCO.
†Nicotiana tabacum L.   TOBACCO.
†Petunia × hybrida Vilm.   GARDEN PETUNIA. (Parentage uncertain; treated by some authors as a part of *P. axillaris* (Lam.) Britton, Sterns & Poggenb.)
*Physalis alkekengi L.   CHINESE-LANTERN.
Physalis heterophylla Nees   CLAMMY GROUND-CHERRY.
†Physalis hispida(Waterf.) Cronquist   PRAIRIE GROUND-CHERRY.   (*P. pumila* Nutt. subsp. *hispida* (Waterf.) W. F. Hinton)
Physalis longifolia Nutt.
    *var. longifolia   LONG-LEAVED GROUND-CHERRY.
    var. subglabrata (Mack. & Bush) Cronquist   SMOOTH GROUND-CHERRY.
†Physalis philadelphica Lam.   TOMATILLO.
Physalis pubescens L.   DOWNY GROUND-CHERRY.
    var. grisea Waterf. (*P. pruinosa* L.)
    var. integrifolia (Dunal) Waterf.
    var. pubescens
Physalis virginiana Mill.   VIRGINIA GROUND-CHERRY.
Solanum carolinense L.   HORSE-NETTLE.
*Solanum dulcamara L.   BITTERSWEET NIGHTSHADE, DEADLY NIGHTSHADE.   (Incl. var. *villosissimum* Desv.)
Solanum nigrum L.   BLACK NIGHTSHADE, COMMON NIGHTSHADE.   (Incl. *S. americanum* Mill. and *S. ptychanthum* Dunal)
†Solanum physalifolium Rusby   HAIRY NIGHTSHADE.   (Treated by some authors as a part of *S. sarrachoides* Sendtn.)
*Solanum rostratum Dunal   BUFFALO-BUR. (Treated by some authors as *S. cornutum* Lam.)
†Solanum triflorum Nutt.   CUT-LEAVED NIGHTSHADE.

†**Solanum tuberosum** L. POTATO, IRISH
POTATO.

■ CONVOLVULACEAE. MORNING-GLORY
FAMILY

†**Calystegia hederacea** Wall. JAPANESE
BINDWEED.
*__Calystegia pubescens__ Lindl. CALIFORNIA-
ROSE. (Treated by some authors as a part
of *C. pellita* (Ledeb.) G. Don)
Calystegia sepium (L.) R. Br. HEDGE
BINDWEED.
Calystegia spithamaea (L.) Pursh UPRIGHT
BINDWEED, LOW BINDWEED.
*__Convolvulus arvensis__ L. COMMON BINDWEED,
FIELD BINDWEED.
*__Ipomoea coccinea__ L. RED MORNING-GLORY.
*__Ipomoea hederacea__ Jacq. IVY-LEAVED
MORNING-GLORY.
Ipomoea lacunosa L. SMALL-FLOWERED
MORNING-GLORY.
Ipomoea pandurata (L.) G. Mey. WILD POTATO-
VINE.
*__Ipomoea purpurea__ (L.) Roth COMMON
MORNING-GLORY.
†**Jacquemontia tamnifolia** (L.) Griseb. TIE-
VINE.

■ CUSCUTACEAE. DODDER FAMILY

Cuscuta cephalanthi Engelm. BUTTONBUSH
DODDER.
Cuscuta compacta Juss. ex Choisy SESSILE
DODDER.
Cuscuta coryli Engelm. HAZEL DODDER.
†**Cuscuta epilinum** Weihe FLAX DODDER.
*__Cuscuta epithymum__ L. CLOVER DODDER.
Cuscuta glomerata Choisy COMPOSITE
DODDER, GLOMERATE DODDER.
Cuscuta gronovii Willd. ex Schult. COMMON
DODDER.
Cuscuta pentagona Engelm. FIVE-ANGLED
DODDER. (Incl. *C. campestris* Yunck.)
Cuscuta polygonorum Engelm. SMARTWEED
DODDER.
†**Cuscuta suaveolens** Ser. SOUTH AMERICAN
DODDER.

■ MENYANTHACEAE. BUCKBEAN FAMILY

Menyanthes trifoliata L. BUCKBEAN, BOGBEAN.
†**Nymphoides peltata** (S. G. Gmel.) Kuntze
YELLOW FLOATING-HEART.

■ POLEMONIACEAE. PHLOX FAMILY

†**Collomia linearis** Nutt. COLLOMIA.
†**Gilia capitata** Sims GILIA.
*__Ipomopsis rubra__ (L.) Wherry STANDING-
CYPRESS. (*Gilia rubra* (L.) A. Heller)
†**Navarretia intertexta** (Benth.) Hook.
NAVARRETIA.
Phlox divaricata L. BLUE PHLOX, WILD SWEET
WILLIAM.
Phlox glaberrima L. **var. triflora** (Michx.)
Reveal & C. R. Broome SMOOTH PHLOX.
(subsp. *triflora* (Michx.) Wherry)
Phlox maculata L. SPOTTED PHLOX, WILD
SWEET WILLIAM. (Incl. var. *purpurea* Michx.)
Phlox ovata L. MOUNTAIN PHLOX, ALLEGHENY
PHLOX. (Incl. *P. latifolia* Michx.)
Phlox paniculata L. GARDEN PHLOX, SWEET
WILLIAM.
Phlox pilosa L. DOWNY PHLOX.
Phlox stolonifera Sims CREEPING PHLOX.
Phlox subulata L. MOSS PHLOX.
    var. brittonii (Small) Wherry (subsp. *brittonii*
    (Small) Wherry; treated by some authors as
    a part of var. *setacea* (L.) Brand)
    var. subulata (subsp. *subulata*)
†**Polemonium caeruleum** L. GARDEN JACOB'S-
LADDER, GREEK VALERIAN.
Polemonium reptans L.
    var. reptans JACOB'S-LADDER.
    var. villosum E. L. Braun BRAUN'S JACOB'S-
    LADDER.

■ HYDROPHYLLACEAE. WATERLEAF
FAMILY

†**Ellisia nyctelea** (L.) L. WATER-POD.
Hydrophyllum appendiculatum Michx.
APPENDAGED WATERLEAF.
Hydrophyllum canadense L. BROAD-LEAVED
WATERLEAF.

Hydrophyllum macrophyllum Nutt. LARGE-LEAVED WATERLEAF.

Hydrophyllum virginianum L. VIRGINIA WATERLEAF.

Phacelia bipinnatifida Michx. FERN-LEAVED SCORPION-WEED.

Phacelia dubia (L.) Trel. SMALL-FLOWERED SCORPION-WEED.

Phacelia purshii Buckley MIAMI-MIST.

Phacelia ranunculacea (Nutt.) Constance WEAK-STEMMED SCORPION-WEED, BLUE SCORPION-WEED.

Order **Lamiales**

■ BORAGINACEAE. BORAGE FAMILY

† **Amsinckia lycopsoides** Lehm. TARWEED.

† **Anchusa azurea** Mill. GARDEN BUGLOSS.

† **Anchusa officinalis** L. COMMON BUGLOSS, COMMON ALKANET.

† **Asperugo procumbens** L. MADWORT.

† **Borago officinalis** L. TALEWORT, COOL-TANKARD.

† **Brunnera macrophylla** (M. Bieb.) I. M. Johnst. SIBERIAN BUGLOSS.

* **Cynoglossum officinale** L. HOUND'S-TONGUE.

Cynoglossum virginianum L. WILD COMFREY.
var. boreale (Fernald) Cooperr. NORTHERN WILD COMFREY. (*C. boreale* Fernald)
var. virginianum SOUTHERN WILD COMFREY.

* **Echium vulgare** L. BLUEWEED, BLUEDEVIL, VIPER'S BUGLOSS. (Incl. *E. pustulatum* Sibth & Sm., *E. vulgare* var. *pustulatum* (Sibth. & Sm.) Coincy)

Hackelia deflexa (Wahlenb.) Opiz **var. americana** (A. Gray) Fernald & I. M. Johnst. NORTHERN STICKSEED. (subsp. *americana* (A. Gray) Á Löve & D. Löve)

Hackelia virginiana (L.) I. M. Johnst. COMMON STICKSEED.

† **Heliotropium europaeum** L. EUROPEAN HELIOTROPE.

† **Heliotropium indicum** L. INDIAN HELIOTROPE.

* **Lappula squarrosa** (Retz.) Dumort. EUROPEAN STICKSEED.

* **Lithospermum arvense** L. CORN GROMWELL. (*Buglossoides arvensis* (L.) I. M. Johnst.)

Lithospermum canescens (Michx.) Lehm. HOARY PUCCOON.

Lithospermum caroliniense (Walter ex J. F. Gmel.) MacMill. **var. croceum** (Fernald) Cronquist PLAINS PUCCOON, HAIRY PUCCOON. (subsp. *croceum* (Fernald) Cusick)

Lithospermum latifolium Michx. AMERICAN GROMWELL.

† **Lithospermum officinale** L. EUROPEAN GROMWELL.

† **Lycopsis arvensis** L. SMALL BUGLOSS. (*Anchusa arvensis* (L.) M. Bieb.)

Mertensia virginica (L.) Pers. ex Link BLUEBELLS, VIRGINIA BLUEBELLS, VIRGINIA COWSLIP.

* **Myosotis arvensis** (L.) Hill FIELD FORGET-ME-NOT, FIELD SCORPION-GRASS.

* **Myosotis discolor** Pers. TWO-COLORED FORGET-ME-NOT. (Incl. M. *versicolor* (Pers.) Sm.)

Myosotis laxa Lehm. SMALLER FORGET-ME-NOT.

Myosotis macrosperma Engelm. BRISTLY SCORPION-GRASS.

* **Myosotis scorpioides** L. TRUE FORGET-ME-NOT.

* **Myosotis stricta** Link ex Roem. & Schult. SMALL-FLOWERED FORGET-ME-NOT. (Treated by some authors as M. *micrantha* Pall. ex Lehm.)

* **Myosotis sylvatica** Ehrh. ex Hoffm. GARDEN FORGET-ME-NOT.

Myosotis verna Nutt. SPRING FORGET-ME-NOT.

Onosmodium molle Michx. **var. hispidissimum** (Mack.) Cronquist FALSE GROMWELL, MARBLE-SEED. (subsp. *hispidissimum* (Mack.) B. Boivin, O. *hispidissimum* Mack.)

* **Symphytum asperum** Lepech. PRICKLY COMFREY.

* **Symphytum officinale** L. COMMON COMFREY.

■ VERBENACEAE. VERBENA FAMILY or VERVAIN FAMILY

Phryma leptostachya L. LOPSEED.

Phyla lanceolata (Michx.) Greene   FROG-FRUIT, FOG-FRUIT.   (*Lippia lanceolata* Michx.)

Verbena bracteata Lag. & Rodr.   PROSTRATE VERVAIN, BRACTED VERVAIN.

*__Verbena canadensis__ (L.) Britton   ROSE VERVAIN.   (*Glandularia canadensis* (L.) Nutt.)

Verbena hastata L.   BLUE VERVAIN.

Verbena simplex Lehm.   NARROW-LEAVED VERVAIN.

Verbena stricta Vent.   HOARY VERVAIN.

Verbena urticifolia L.   WHITE VERVAIN.

Verbena × engelmannii Moldenke (**Verbena hastata × V. urticifolia**)   ENGELMANN'S VERVAIN.

Verbena × moechina Moldenke (**Verbena simplex × V. stricta**)   MOLDENKE'S VERVAIN.

†**Vitex negundo** L. **var. heterophylla** (Franch.) Rehder   CUT-LEAVED CHASTE-TREE.

■ LAMIACEAE or LABIATAE.   MINT FAMILY

†**Acinos arvensis** (Lam.) Dandy   MOTHER-OF-THYME, BASIL-THYME.   (*Satureja acinos* (L.) Scheele)

Agastache nepetoides (L.) Kuntze   YELLOW GIANT-HYSSOP.

Agastache scrophulariifolia (Willd.) Kuntze   PURPLE GIANT-HYSSOP.

†**Ajuga genevensis** L.   ERECT BUGLEWEED.

*__Ajuga reptans__ L.   CARPET BUGLEWEED.

†**Ballota nigra** L. **var. alba** (L.) Sm.   BLACK HOREHOUND.

Blephilia ciliata (L.) Benth.   DOWNY WOODMINT.

Blephilia hirsuta (Pursh) Benth.   HAIRY WOODMINT.

Calamintha arkansana (Nutt.) Shinners   LIMESTONE SAVORY, LOW CALAMINT.   (*Satureja arkansana* (Nutt.) Briq., *S. glabella* (Michx.) Briq. var. *angustifolia* (Torr.) Svenson)

Clinopodium vulgare L.   WILD BASIL.   (*Satureja vulgaris* (L.) Fritsch)

Collinsonia canadensis L.   RICHWEED, STONEROOT.

Collinsonia verticillata Baldwin   EARLY STONEROOT.

Cunila origanoides (L.) Britton   COMMON DITTANY.

†**Dracocephalum parviflorum** Nutt.   AMERICAN DRAGONHEAD.

†**Galeopsis ladanum** L.   RED HEMP-NETTLE.

*__Galeopsis tetrahit__ L.   COMMON HEMP-NETTLE.

*__Glechoma hederacea__ L.   GROUND IVY, GILL-OVER-THE-GROUND.   (Incl. var. *micrantha* Moric.)

Hedeoma hispida Pursh   ROUGH PENNYROYAL, ROUGH MOCK PENNYROYAL.

Hedeoma pulegioides (L.) Pers.   AMERICAN PENNYROYAL, MOCK PENNYROYAL.

Isanthus brachiatus (L.) Britton, Sterns & Poggenb.   FALSE PENNYROYAL.   (*Trichostema brachiatum* L.)

*__Lamium amplexicaule__ L.   HENBIT.

†**Lamium maculatum** L.   SPOTTED DEADNETTLE, SPOTTED HENBIT.

*__Lamium purpureum__ L.   PURPLE DEAD-NETTLE.

*__Leonurus cardiaca__ L.   COMMON MOTHERWORT.

*__Leonurus marrubiastrum__ L.   HOREHOUND MOTHERWORT.

†**Leonurus sibiricus** L.   HONEYWEED, ASIAN MOTHERWORT.

Lycopus americanus Muhl. ex W. P. C. Barton   CUT-LEAVED WATER-HOREHOUND.

*__Lycopus asper__ Greene   ROUGH WATER-HOREHOUND.

*__Lycopus europaeus__ L.   EUROPEAN BUGLEWEED.

Lycopus rubellus Moench   STALKED WATER-HOREHOUND.

Lycopus virginicus L.   VIRGINIA WATER-HOREHOUND, VIRGINIA BUGLEWEED.   (Incl. *L.* × *sherardii* Steele)
var. pauciflorus Benth. (*L. uniflorus* Michx.)
var. virginicus

†**Marrubium vulgare** L.   COMMON HOREHOUND, WHITE HOREHOUND.

Meehania cordata (Nutt.) Britton   MEEHANIA.

*__Melissa officinalis__ L.   COMMON BALM.

Mentha arvensis L. **var. canadensis** (L.) Kuntze   FIELD MINT.   (Incl. var. *glabrata* (Benth.) Fernald and *M. gentilis* L.; *M. canadensis* L.)

*__Mentha longifolia__ (L.) L.   EUROPEAN HORSEMINT.

*Mentha spicata L. SPEARMINT.

*Mentha × gracilis Sole (Mentha arvensis ×
   M. spicata) RED MINT. (Incl. M. cardiaca
   (Gray) J. Gerard ex Baker and M. gentilis of
   authors, not L.)

*Mentha × piperita L. (Mentha aquatica L. ×
   M. spicata)
   *var. citrata (Ehrh.) Boivin BERGAMOT
      MINT. (subsp. citrata (Ehrh.) Briq.,
      M. citrata Ehrh.; treated by some authors
      as a part of M. aquatica L.)
   *var. piperita PEPPERMINT. (subsp. piperita)

†Mentha × rotundifolia (L.) Huds. (Mentha
   longifolia × M. suaveolens Ehrh.) APPLE
   MINT.

†Mentha × villosa Huds. var. alopecuroides
   (Hull) Briq. (Mentha spicata × M. suaveolens
   Ehrh.) FOXTAIL MINT. (M. alopecuroides
   Hull)

Monarda didyma L. OSWEGO-TEA, BEE-BALM.

Monarda fistulosa L. WILD BERGAMOT.
   var. clinopodia (L.) Cooperr. (M. clinopodia L.)
   var. fistulosa

Monarda punctata L. var. villicaulis (Pennell)
   E. J. Palmer & Steyerm. DOTTED HORSEMINT.

Monarda × media Willd. (Monarda didyma ×
   M. fistulosa) PURPLE BERGAMOT.

*Nepeta cataria L. CATNIP.

*Origanum vulgare L. MARJORAM, POT
   MARJORAM, OREGANO.

*Perilla frutescens (L.) Britton
   BEEFSTEAK-PLANT. (Incl. var. crispa
   (Benth.) W. Deane)

Physostegia virginiana (L.) Benth. FALSE
   DRAGONHEAD, OBEDIENCE, OBEDIENT-PLANT.
   var. arenaria Shimek (subsp. praemorsa
      (Shinners) P. D. Cantino)
   var. virginiana (subsp. virginiana)

Prunella vulgaris L. SELF-HEAL, HEAL-ALL.
   (Incl. var. lanceolata (W. Bartram) Fernald)

Pycnanthemum incanum (L.) Michx. HOARY
   MOUNTAIN-MINT. (Incl. var. puberulum
   (E. Grant & Epling) Fernald)

Pycnanthemum muticum (Michx.) Pers.
   BLUNT MOUNTAIN-MINT.

Pycnanthemum pycnanthemoides (Leavenw.)

Fernald SOUTHERN MOUNTAIN-MINT.
   (Incl. var. viridifolium Fernald)

Pycnanthemum tenuifolium Schrad. NARROW-
   LEAVED MOUNTAIN-MINT.

Pycnanthemum verticillatum (Michx.) Pers.
   var. pilosum (Nutt.) Cooperr. HAIRY
      MOUNTAIN-MINT. (P. pilosum Nutt.)
   var. verticillatum VERTICILLATE MOUNTAIN-
      MINT.

Pycnanthemum virginianum (L.) T. Durand &
   B. D. Jacks. ex B. L. Rob. & Fernald VIRGINIA
   MOUNTAIN-MINT.

†Salvia azurea Michx. ex Lam. var. grandiflora
   Benth. BLUE SAGE.

†Salvia coccinea Buc'hoz ex Etl. CRIMSON
   SAGE, TEXAS SAGE.

†Salvia farinacea Benth. MEALY-CUP SAGE.

Salvia lyrata L. LYRE-LEAVED SAGE.

†Salvia officinalis L. COMMON SAGE, GARDEN
   SAGE.

†Salvia pratensis L. MEADOW SAGE.

Salvia reflexa Hornem. ROCKY MOUNTAIN
   SAGE.

†Salvia splendens Sellow ex Roem. & Schult.
   SCARLET SAGE.

†Salvia × superba Stapf (Salvia × sylvestris L. ×
   S. villicaulis Borbás) SHOWY SAGE.

†Satureja hortensis L. SAVORY, SUMMER
   SAVORY.

Scutellaria elliptica Muhl. ex Spreng. HAIRY
   SKULLCAP.
   var. elliptica
   var. hirsuta (Short & R. Peter) Fernald

Scutellaria epilobiifolia A. Ham. MARSH
   SKULLCAP. (Treated by some authors as
   a part of S. galericulata L.)

Scutellaria incana Biehler DOWNY SKULLCAP.

Scutellaria integrifolia L. HYSSOP SKULLCAP.

Scutellaria lateriflora L. MAD-DOG SKULLCAP.

Scutellaria nervosa Pursh VEINED SKULLCAP.
   var. calvifolia Fernald
   var. nervosa

Scutellaria ovata Hill var. versicolor (Nutt.)
   Fernald HEART-LEAVED SKULLCAP. (subsp.
   versicolor (Nutt.) Epling)

Scutellaria parvula Michx. SMALL SKULLCAP.

var. **leonardii** (Epling) Fernald (*S. leonardii* Epling)

var. **parvula**

Scutellaria saxatilis Riddell   ROCK SKULLCAP.

Scutellaria serrata Andrews   SHOWY SKULLCAP.

Stachys aspera Michx.   ROUGH HEDGE-NETTLE. (*S. hyssopifolia* Michx. var. *ambigua* A. Gray)

†**Stachys germanica** L.   DOWNY WOUNDWORT.

Stachys nuttallii Shuttlew. ex Benth.   HEART-LEAVED HEDGE-NETTLE.

Stachys palustris L.   MARSH HEDGE-NETTLE, MARSH WOUNDWORT.   (Incl. var. *homotricha* Fernald and *S. pilosa* Nutt. var. *pilosa* and var. *arenicola* (Britton) G. A. Mulligan & D. B. Munro)

†**Stachys sylvatica** L.   WOODLAND HEDGE-NETTLE, WHITE-SPOT.

Stachys tenuifolia Willd.   COMMON HEDGE-NETTLE.   (Incl. *S. hispida* Pursh)

Synandra hispidula (Michx.) Baill.   SYNANDRA.

Teucrium canadense L.   AMERICAN GERMANDER.

var. **canadense** (subsp. *canadense*; incl. var. *virginicum* (L.) Eaton)

var. **occidentale** (A. Gray) E. M. McClint. & Epling (subsp. *occidentale* (A. Gray) W. A. Weber)

*__**Thymus pulegioides** L.__   CREEPING THYME, WILD THYME.   (Until recently, Ohio plants of this genus were generally assigned to *T. serpyllum* L.)

Trichostema dichotomum L.

var. **dichotomum**   BLUECURLS, BASTARD PENNYROYAL.

var. **lineare** (Walter) Pursh   NARROW-LEAVED BLUECURLS.   (*T. setaceum* Houtt.)

Order **Callitrichales**

■ CALLITRICHACEAE.   WATER-STARWORT FAMILY

Callitriche heterophylla Pursh   LARGER WATER-STARWORT.

Callitriche palustris L.   VERNAL WATER-STARWORT.   (Incl. *C. verna* L.)

Callitriche terrestris Raf.   TERRESTRIAL WATER-STARWORT.

Order **Plantaginales**

■ PLANTAGINACEAE.   PLANTAIN FAMILY

*__**Plantago aristata** Michx.__   BRACTED PLANTAIN.

Plantago cordata Lam.   HEART-LEAVED PLANTAIN.

*__**Plantago lanceolata** L.__   ENGLISH PLANTAIN, BUCKHORN PLANTAIN, NARROW-LEAVED PLANTAIN.

*__**Plantago major** L.__   COMMON PLANTAIN.

Plantago patagonica Jacq.   SALT-AND-PEPPER-PLANT, WOOLLY PLANTAIN.

*__**Plantago psyllium** L.__   SAND PLANTAIN, FLEAWORT.

Plantago rugelii Decne.   RUGEL'S PLANTAIN.

Plantago virginica L.   DWARF PLANTAIN, HOARY PLANTAIN.

Order **Scrophulariales**

■ BUDDLEJACEAE.   BUTTERFLY-BUSH FAMILY

†**Buddleja davidii** Franch.   BUTTERFLY-BUSH.

■ OLEACEAE.   OLIVE FAMILY

Chionanthus virginicus L.   FRINGE-TREE.

†**Fontanesia fortunei** Carrière   FALSE PRIVET, FONTANESIA.

†**Forsythia × intermedia** Zabel (**Forsythia suspensa** (Thunb.) Vahl × **F. viridissima** Lindl.) FORSYTHIA, GOLDEN-BELLS.

Fraxinus americana L.

var. **americana**   WHITE ASH.

var. **biltmoreana** (Beadle) J. Wright ex Fernald   BILTMORE ASH.

Fraxinus nigra Marshall   BLACK ASH.

Fraxinus pennsylvanica Marshall

var. **pennsylvanica**   RED ASH.

var. **subintegerrima** (M. Vahl) Fernald   GREEN ASH.

Fraxinus profunda (Bush) Bush   PUMPKIN ASH. (Treated by some authors as *F. tomentosa* F. Michx.)

Fraxinus quadrangulata Michx.   BLUE ASH.

*Ligustrum obtusifolium Siebold & Zucc.
JAPANESE PRIVET.

†Ligustrum ovalifolium Hassk.   CALIFORNIA
PRIVET.

*Ligustrum vulgare L.   COMMON PRIVET.

†Syringa vulgaris L.   LILAC.

■ SCROPHULARIACEAE.   FIGWORT
FAMILY

Agalinis purpurea (L.) Pennell   PURPLE
AGALINIS, PURPLE-FOXGLOVE.
    var. parviflora (Benth.) B. Boivin   SMALL
    PURPLE AGALINIS, SMALL PURPLE-
    FOXGLOVE.   (A. paupercula (A. Gray)
    Britton)
    var. purpurea   LARGE PURPLE AGALINIS,
    LARGE PURPLE-FOXGLOVE.

Agalinis skinneriana (A. W. Wood) Britton
SKINNER'S AGALINIS, SKINNER'S-FOXGLOVE.
(Incl. A. gattingeri (Small) Small ex Britton—
GATTINGER'S-FOXGLOVE)

Agalinis tenuifolia (Vahl) Raf.   SLENDER
AGALINIS, SLENDER-FOXGLOVE.
    var. macrophylla (Benth.) S. F. Blake
    var. parviflora (Nutt.) Pennell
    var. tenuifolia

†Antirrhinum majus L.   SNAPDRAGON,
GARDEN SNAPDRAGON.

†Antirrhinum orontium L.   LESSER
SNAPDRAGON.   (Misopates orontium (L.) Raf.)

Aureolaria flava (L.) Farw.   SMOOTH FALSE
FOXGLOVE.
    var. flava (subsp. flava)
    var. macrantha Pennell (subsp. macrantha
    (Pennell) Pennell)

Aureolaria laevigata (Raf.) Raf.   ENTIRE-LEAVED
FALSE FOXGLOVE.

Aureolaria pedicularia (L.) Raf.   FERN-LEAVED
FALSE FOXGLOVE.
    var. ambigens (Fernald) Farw.   PRAIRIE FERN-
    LEAVED FALSE FOXGLOVE.   (subsp. ambigens
    (Fernald) Pennell)
    var. pedicularia   WOODLAND FERN-LEAVED
    FALSE FOXGLOVE.   (subsp. pedicularia)

Aureolaria virginica (L.) Pennell   DOWNY
FALSE FOXGLOVE.

Besseya bullii (Eaton) Rydb.   KITTEN-TAILS,
BESSEYA.

Buchnera americana L.   BLUEHEARTS,
AMERICAN BLUEHEARTS.

Castilleja coccinea (L.) Spreng.   INDIAN
PAINTBRUSH, PAINTED-CUP.

*Chaenorrhinum minus (L.) Lange   DWARF
SNAPDRAGON, LESSER TOADFLAX.

Chelone glabra L.   TURTLEHEAD.
    var. elatior T. A. Frick (subsp. elatior (T. A.
    Frick) Pennell)
    var. glabra (subsp. glabra; incl. var. elongata
    Pennell & Wherry)
    var. linifolia N. Coleman (subsp. linifolia
    (N. Coleman) Pennell)

Collinsia verna Nutt.   BLUE-EYED MARY.

*Cymbalaria muralis P. Gaertn., B. Mey. &
Scherb.   KENILWORTH IVY.

Dasistoma macrophylla (Nutt.) Raf.   MULLEIN-
FOXGLOVE.

†Digitalis grandiflora Mill.   YELLOW FOXGLOVE.

†Digitalis lanata Ehrh.   GRECIAN FOXGLOVE.

†Digitalis lutea L.   STRAW FOXGLOVE.

†Digitalis purpurea L.   COMMON FOXGLOVE.

Gratiola neglecta Torr.   COMMON HEDGE-
HYSSOP.

Gratiola virginiana L.   ROUND-FRUITED HEDGE-
HYSSOP.

Gratiola viscidula Pennell   VISCID HEDGE-
HYSSOP, SHORT'S HEDGE-HYSSOP.   (Incl. var.
shortii (Durand ex Pennell) Gleason)

*Kickxia elatine (L.) Dumort.   SHARP-POINTED
CANCERWORT.

*Kickxia spuria (L.) Dumort.   ROUND-LEAVED
CANCERWORT.

Leucospora multifida (Michx.) Nutt.
LEUCOSPORA.

Linaria canadensis (L.) Chaz.   OLD-FIELD
TOADFLAX, BLUE TOADFLAX.   (Nuttallanthus
canadensis (L.) D. A. Sutton)

†Linaria dalmatica (L.) Mill.   DALMATIAN
TOADFLAX.   (L. genistifolia subsp. dalmatica (L.) Marie & Petitm.)

†Linaria genistifolia (L.) Mill.   EUROPEAN
TOADFLAX.

*Linaria vulgaris Mill. Butter-and-eggs,
Yellow Toadflax.

Lindernia dubia (L.) Pennell False Pimpernel.
var. anagallidea (Michx.) Cooperr. (L. anagallidea (Michx.) Pennell)
var. dubia

*Mazus pumilus (Burm. f.) Steenis Mazus.

Melampyrum lineare Desr. Cow-wheat.
var. americanum (Michx.) Beauverd
var. latifolium Barton

Mimulus alatus Aiton Sharp-winged
Monkey-flower.

Mimulus ringens L. Common Monkey-
flower.

Pedicularis canadensis L. Common Louse-
wort, Wood-betony.

Pedicularis lanceolata Michx. Swamp Louse-
wort.

Penstemon calycosus Small Long-sepaled
Beard-tongue. (P. laevigatus subsp. calycosus (Small) R. W. Benn.)

Penstemon canescens (Britton) Britton Gray
Beard-tongue.

†Penstemon cobaea Nutt. Cobaean Beard-
tongue.

Penstemon digitalis Nutt. ex Sims Foxglove
Beard-tongue. (Incl. P. alluviorum Pennell)

†Penstemon grandiflorus Nutt. Large-
flowered Beard-tongue.

Penstemon hirsutus (L.) Willd. Hairy Beard-
tongue.

Penstemon laevigatus Aiton Smooth Beard-
tongue.

Penstemon pallidus Small Downy White
Beard-tongue.

Penstemon tubaeflorus Nutt. White-wand
Beard-tongue.

Scrophularia lanceolata Pursh Early Figwort,
Lance-leaved Figwort.

Scrophularia marilandica L. Maryland
Figwort, Carpenter's-square.

Tomanthera auriculata (Michx.) Raf. Ear-
leaved-foxglove. (Agalinis auriculata
(Michx.) S. F. Blake)

*Verbascum blattaria L. Moth Mullein.
*forma blattaria Yellow-flowered Moth
Mullein.

*forma erubescens Brügger White-
flowered Moth Mullein.

†Verbascum phlomoides L. Clasping-leaved
Mullein.

†Verbascum phoeniceum L. Purple
Mullein.

*Verbascum thapsus L. Common Mullein,
Velvet-plant.

†Verbascum virgatum Stokes Twiggy
Mullein.

†Veronica agrestis L. Field Speedwell.

Veronica americana Schwein. ex Benth.
American Brooklime.

Veronica anagallis-aquatica L. Water
Speedwell.

*Veronica arvensis L. Corn Speedwell.

*Veronica beccabunga L. European
Brooklime.

Veronica catenata Pennell Sweet-water
Speedwell.

*Veronica chamaedrys L. Germander
Speedwell, Bird's-eye Speedwell.

*Veronica filiformis Sm. Blue-eyed
Speedwell, Slender Speedwell.

*Veronica hederifolia L. Ivy-leaved
Speedwell.

†Veronica latifolia L. Broad-leaved
Speedwell. (Incl. V. austriaca L. subsp.
teucrium (L.) D. A. Webb, V. teucrium L.)

†Veronica longifolia L. Garden Speedwell.

*Veronica officinalis L. Common
Speedwell.

Veronica peregrina L.
var. peregrina Purslane Speedwell.
(subsp. peregrina)
*var. xalapensis (Kunth) H. St. John & F. A.
Warren Hairy Purslane Speedwell.
(subsp. xalapensis (Kunth) Pennell)

*Veronica persica Poir. Cat's-eye Speedwell,
Bird's-eye Speedwell.

*Veronica polita Fr. Wayside Speedwell.

Veronica scutellata L. Marsh Speedwell.

*Veronica serpyllifolia L. Thyme-leaved
Speedwell.

†Veronica verna L. Spring Speedwell.

Veronicastrum virginicum (L.) Farw.
Culver's-root.

- OROBANCHACEAE. Broom-rape Family

Conopholis americana (L.) Wallr. Squaw-
root.
Epifagus virginiana (L.) Barton Beech-drops.
Orobanche ludoviciana Nutt. Louisiana
Broom-rape.
Orobanche uniflora L. One-flowered Broom-
rape, One-flowered Cancer-root.

- ACANTHACEAE. Acanthus Family

Justicia americana (L.) M. Vahl Water-
willow.
Ruellia caroliniensis (J. F. Gmel.) Steud.
Carolina Ruellia.
Ruellia humilis Nutt. **var. calvescens** Fernald
Hairy Ruellia, Wild Petunia.
Ruellia strepens L. Smooth Ruellia.
Ruellia caroliniensis × **R. humilis**
Ruellia humilis × **R. strepens**

- PEDALIACEAE. Sesame Family

†**Proboscidea louisianica** (Mill.) Thell.
Unicorn-plant, Proboscis-flower.

- BIGNONIACEAE. Trumpet-creeper
Family

Bignonia capreolata L. Cross-vine.
Campsis radicans (L.) Seem. ex Bureau
Trumpet-creeper, Trumpet-vine.
*Catalpa bignonioides** Walter Southern
Catalpa, Indian-bean.
*Catalpa ovata** G. Don Chinese Catalpa.
*Catalpa speciosa** (Warder) Warder ex Engelm.
Northern Catalpa, Cigar-tree.
*Paulownia tomentosa** (Thunb.) Siebold &
Zucc. ex Steud. Princess-tree, Royal
Paulownia.

- LENTIBULARIACEAE. Bladderwort
Family

Utricularia cornuta Michx. Horned
Bladderwort.

Utricularia geminiscapa Benj. Two-scaped
Bladderwort.
Utricularia gibba L. Humped Bladderwort.
Utricularia intermedia Hayne Flat-leaved
Bladderwort.
Utricularia minor L. Lesser Bladderwort.
Utricularia vulgaris L. Greater
Bladderwort. (Incl. *U. macrorhiza*
Leconte)

Order **Campanulales**

- CAMPANULACEAE. Bellflower Family

Campanula americana L. Tall Bellflower.
(*Campanulastrum americanum* (L.) Small)
Campanula aparinoides Pursh Marsh
Bellflower.
  var. aparinoides
  var. grandiflora Holz. (*C. uliginosa* Rydb.)
*Campanula rapunculoides** L. European
Bellflower, Creeping Bellflower.
Campanula rotundifolia L. Harebell.
†**Campanula trachelium** L. Nettle-leaved
Bellflower.
Lobelia cardinalis L. Cardinal-flower.
Lobelia inflata L. Indian-tobacco.
Lobelia kalmii L. Kalm's Lobelia.
Lobelia puberula Michx. Downy Lobelia.
  (Incl. var. *simulans* Fernald)
Lobelia siphilitica L. Great Lobelia, Blue
Lobelia, Great Blue Lobelia.
Lobelia spicata Lam. Pale-spiked Lobelia.
  var. leptostachys (A. DC.) Mack. & Bush
      (Incl. var. *campanulata* McVaugh)
  var. spicata
Triodanis perfoliata (L.) Nieuwl. Venus'-
looking-glass.

Order **Rubiales**

- RUBIACEAE. Madder Family

Cephalanthus occidentalis L. Buttonbush.
  (Incl. var. *pubescens* Raf.)
Diodia teres Walter Rough Buttonweed.
Diodia virginiana L. Southern
Buttonweed.

Galium aparine L.　Cleavers.　(Incl. G.
　spurium L.)
Galium asprellum Michx.　Rough Bedstraw.
Galium boreale L.　Northern Bedstraw.
Galium circaezans Michx.　Wild Licorice.
　(Incl. var. *hypomalacum* Fernald)
Galium concinnum Torr. & A. Gray　Shining
　Bedstraw.
Galium labradoricum (Wiegand) Wiegand　Bog
　Bedstraw.
Galium lanceolatum Torr.　Lance-leaved
　Bedstraw, Lance-leaved Wild Licorice.
*Galium mollugo L.　White Bedstraw.
Galium obtusum Bigelow　Stiff Marsh
　Bedstraw.
*Galium odoratum (L.) Scop.　Sweet
　Woodruff.
Galium palustre L.　Northern Marsh
　Bedstraw, Marsh Bedstraw.
*Galium pedemontanum (Bellardi) All.
　Piedmont Bedstraw.　(*Cruciata pedemontana*
　(Bellardi) Ehrend.)
Galium pilosum Aiton　Hairy Bedstraw.
Galium tinctorium (L.) Scop.　Southern
　Three-lobed Bedstraw.
Galium trifidum L.　Small Bedstraw.
Galium triflorum Michx.　Sweet-scented
　Bedstraw, Fragrant Bedstraw.
*Galium verum L.　Yellow Bedstraw.
Hedyotis nigricans (Lam.) Fosberg　Narrow-
　leaved Summer Bluets.　(*Houstonia nigri-
　cans* (Lam.) Fernald)
Houstonia caerulea L.　Bluets, Quaker
　Ladies, Innocence.　(*Hedyotis caerulea* (L.)
　Hook.)
Houstonia canadensis Willd. ex Roem. &
　Schult.　Canada Bluets, Fringed
　Houstonia.　(Incl. *H. setiscaphia* L. G.
　Carr; *Hedyotis canadensis* (Willd. ex Roem &
　Schult.) Fosberg)
Houstonia longifolia Gaertn.　Long-leaved
　Summer Bluets, Long-leaved Houstonia.
　(*Hedyotis longifolia* (Gaertn.) Hook.)
Houstonia purpurea L.　Large Summer
　Bluets, Large Houstonia.　(Incl. var. *caly-
　cosa* A. Gray and *H. lanceolata* (Poir.) Britton;
　Hedyotis purpurea (L.) Torr. & A. Gray)

Houstonia longifolia × H. purpurea
Mitchella repens L.　Partridge-berry.
*Sherardia arvensis L.　Field Madder.
Spermacoce glabra Michx.　Smooth
　Buttonweed.

Order **Dipsacales**

■ CAPRIFOLIACEAE.　Honeysuckle Family

Diervilla lonicera Mill.　Bush-honeysuckle.
Linnaea borealis L. var. americana (J. Forbes)
　Rehder　Twinflower, American
　Twinflower.　(subsp. *americana* (J. Forbes)
　Hultén ex R. T. Clausen; var. *americana* is
　treated by some authors as a part of var. *longi-
　flora* Torr. or subsp. *longiflora* (Torr.) Hultén)
Lonicera canadensis Barton ex Marshall
　Canada Fly Honeysuckle.
Lonicera dioica L.　Wild Honeysuckle.　(Incl.
　var. *dasygyna* (Rehder) Gleason and var. *glau-
　cescens* (Rydb.) Butters)
Lonicera flava Sims var. flavescens Gleason
　Yellow Honeysuckle, Pale Yellow
　Honeysuckle.　(Incl. *L. flavida* Cockerell
　ex Rehder)
*Lonicera fragrantissima Lindl. & Paxt.
　Winter Honeysuckle, Sweet-breath-of-
　spring.
*Lonicera japonica Thunb.　Japanese
　Honeysuckle.
*Lonicera maackii (Rupr.) Maxim.　Amur
　Honeysuckle.
*Lonicera morrowii A. Gray　Morrow's
　Honeysuckle.
Lonicera oblongifolia (Goldie) Hook.　Swamp
　Fly Honeysuckle.
Lonicera reticulata Raf.　Grape Honeysuckle.
　(Incl. *L. prolifera* (G. Kirchn.) Booth ex
　Rehder)
*Lonicera sempervirens L.　Trumpet
　Honeysuckle.
*Lonicera tatarica L.　Tatarian Honeysuckle.
Lonicera villosa (Michx.) Schult.　Mountain
　Fly Honeysuckle.　(Incl. var. *tonsa* Fernald;
　L. caerulea var. *villosa* (Michx.) Torr. &
　A. Gray)

*Lonicera xylosteum L.   EUROPEAN FLY
 HONEYSUCKLE.
*Lonicera × bella Zabel (**Lonicera morrowii** ×
 L. tatarica)   SHOWY PINK HONEYSUCKLE.
*Lonicera × xylosteoides Tausch (**Lonicera
 tatarica** × **L. xylosteum**)   GARDEN
 HONEYSUCKLE.
Sambucus canadensis L.   ELDERBERRY, COMMON
 ELDER.
Sambucus pubens Michx.   RED-BERRIED ELDER.
 (*S. racemosa* L. var. *pubens* (Michx.) Koehne)
Symphoricarpos albus (L.) S. F. Blake
 SNOWBERRY.
 var. albus (subsp. *albus*)
 *var. laevigatus (Fernald) S. F. Blake (subsp.
  laevigatus (Fernald) Hultén)
*Symphoricarpos occidentalis Hook.
 WOLFBERRY.
Symphoricarpos orbiculatus Moench
 CORALBERRY, INDIAN-CURRANT.
Triosteum angustifolium L.   YELLOW HORSE-
 GENTIAN.
Triosteum aurantiacum E. P. Bicknell   WILD
 COFFEE.   (Incl. var. *glaucescens* Wiegand
 and var. *illinoense* (Wiegand) E. J. Palmer &
 Steyerm.)
Triosteum perfoliatum L.   TINKER'S-WEED.
Viburnum acerifolium L.   MAPLE-LEAVED
 VIBURNUM, MAPLE-LEAVED ARROW-WOOD.
 (Incl. var. *glabrescens* Rehder)
Viburnum alnifolium Marshall   HOBBLEBUSH.
 (Incl. *V. lantanoides* Michx.)
† Viburnum buddleifolium C. H. Wright
 BUDDLEJA VIBURNUM, CHINESE VIBURNUM.
Viburnum cassinoides L.   WITHE-ROD, WILD
 RAISIN.   (*V. nudum* L. var. *cassinoides* (L.)
 Torr. & A. Gray)
Viburnum dentatum L.   SOUTHERN ARROW-
 WOOD.   (Incl. var. *deamii* (Rehder) Fernald,
 var. *indianense* (Rehder) Gleason, var. *sca-
 brellum* Torr. & A. Gray, and var. *venosum*
 (Britton) Gleason)
*Viburnum lantana L.   WAYFARING-TREE.
Viburnum lentago L.   NANNYBERRY,
 SHEEPBERRY.
Viburnum molle Michx.   SOFT-LEAVED ARROW-
 WOOD.

Viburnum opulus L.
 var. americanum Aiton   HIGHBUSH-
  CRANBERRY.   (subsp. *trilobum* (Marshall)
  R. T. Clausen, *V. trilobum* Marshall)
 *var. opulus   EUROPEAN CRANBERRY-BUSH,
  GUELDER-ROSE.   (subsp. *opulus*)
*Viburnum plicatum Thunb.   JAPANESE
 SNOWBALL, DOUBLE-FILE VIBURNUM.
Viburnum prunifolium L.   BLACK-HAW.
Viburnum rafinesquianum Schult.   DOWNY
 ARROW-WOOD.
 var. affine (Bush ex C. K. Schneid.) House
 var. rafinesquianum
Viburnum recognitum Fernald   NORTHERN
 ARROW-WOOD.   (*V. dentatum* var. *lucidulum*
 Aiton)
Viburnum rufidulum Raf.   SOUTHERN BLACK-
 HAW.
*Viburnum × rhytidophylloides J. V. Suringar
 (**Viburnum lantana** × **V. rhytidophyllum**
 Hemsl.)   LANTANA-LEAVED VIBURNUM.

■ VALERIANACEAE.   VALERIAN FAMILY

Valeriana edulis Nutt. ex Torr. & A. Gray **var.
 ciliata** (Torr. & A. Gray) Cronquist   PRAIRIE
 VALERIAN.   (subsp. *ciliata* (Torr. & A. Gray)
 F. G. Mey., *V. ciliata* Torr. & A. Gray)
*Valeriana officinalis L.   GARDEN VALERIAN.
Valeriana pauciflora Michx.   LARGE-FLOWERED
 VALERIAN.
Valeriana uliginosa (Torr. & A. Gray) Rydb.
 SWAMP VALERIAN.
Valerianella chenopodiifolia (Pursh) DC.
 GOOSEFOOT CORN-SALAD.
*Valerianella locusta (L.) Latourr.   EUROPEAN
 CORN-SALAD, LAMB'S-LETTUCE.
† Valerianella radiata (L.) Dufr.   AMERICAN
 CORN-SALAD.
Valerianella umbilicata (Sull.) A. W. Wood
 BEAKED CORN-SALAD.

■ DIPSACACEAE.   TEASEL FAMILY

*Dipsacus fullonum L.   COMMON TEASEL, WILD
 TEASEL.   (Incl. *D. sylvestris* Huds.)
*Dipsacus laciniatus L.   CUT-LEAVED TEASEL.

† **Dipsacus sativus** (L.) Honck.   Fuller's
Teasel.
† **Scabiosa columbaria** L.   Small Scabious.

Order **Asterales**

■ ASTERACEAE or COMPOSITAE.
Sunflower Family or Aster Family

Achillea millefolium L.   Yarrow.
var. millefolium (subsp. *millefolium*)
   var. occidentalis DC. (subsp. *occidentalis* (DC.)
      Hyl.; incl. subsp. *lanulosa* (Nutt.) Piper,
      A. lanulosa Nutt.)
Achillea ptarmica L.   Sneezewort,
   Sneezeweed.
Ageratina altissima (L.) R. M. King & H. Rob.
   White Snakeroot.   (*Eupatorium rugosum*
   Houtt.)
Ageratina aromatica (L.) Spach   Small
   White Snakeroot, Small-leaved White
   Snakeroot.   (*Eupatorium aromaticum* L.)
Ambrosia artemisiifolia L.   Common Ragweed.
Ambrosia bidentata Michx.   Lance-leaved
   Ragweed.
† **Ambrosia psilostachya** DC.   Western
   Ragweed.
Ambrosia trifida L.   Giant Ragweed.
Ambrosia × helenae Rouleau (**Ambrosia arte-**
   misiifolia × A. trifida)   Intermediate
   Ragweed.
† **Amphiachyris dracunculoides** (DC.) Nutt.
   Broomweed, Broom-snakeroot.   (*Gutier-*
   rezia dracunculoides (DC.) S. F. Blake)
Anaphalis margaritacea (L.) Benth. & Hook.
   Pearly Everlasting.
Antennaria neglecta Greene   Field Pussy-toes,
   Field Everlasting.
   var. canadensis (Greene) Cronquist (*A. ho-*
      wellii Greene subsp. *canadensis* (Greene)
      R. J. Bayer; *A. neodioica* Greene subsp.
      canadensis (Greene) R. J. Bayer & Stebbins)
   var. neglecta
   var. neodioica (Greene) Cronquist (*A. ho-*
      wellii Greene subsp. *neodioica* (Greene) R. J.
      Bayer; *A. neodioica* Greene subsp. *neodioica*)
   var. petaloidea (Fernald) Cronquist (*A. ho-*
      wellii Greene subsp. *petaloidea* (Fernald) R. J.

Bayer; *A. neodioica* Greene subsp. *petaloidea*
   (Fernald) R. J. Bayer & Stebbins)
Antennaria plantaginifolia (L.) Richardson
   Plantain-leaved Pussy-toes, Plantain-
   leaved Everlasting.
   var. ambigens (Greene) Cronquist (*A. parlinii*
      Fernald subsp. *falax* (Greene) R. J. Bayer &
      Stebbins)
   var. parlinii (Fernald) Cronquist (*A. parlinii*
      Fernald subsp. *parlinii*)
   var. plantaginifolia
Antennaria solitaria Rydb.   Single-headed
   Pussy-toes, Single-headed Everlasting.
Antennaria virginica Stebbins   Shale Barren
   Pussy-toes, Shale Barren Everlasting.
Anthemis arvensis L.   Corn Cotula, Corn
   Chamomile.
Anthemis cotula L.   Mayweed, Stinking
   Cotula, Stinking Chamomile.
† **Anthemis nobilis** L.   Garden Chamomile,
   Russian Chamomile.
Anthemis tinctoria L.   Yellow Chamomile,
   Golden Marguerite.
Arctium lappa L.   Great Burdock.
Arctium minus Bernh.   Common Burdock.
Arnoglossum atriplicifolium (L.) H. Rob.   Pale
   Indian-plantain.   (*Cacalia atriplicifolia* L.)
Arnoglossum muhlenbergii (Sch. Bip.) H. Rob.
   Great Indian-plantain.   (*Cacalia muhlen-*
   bergii (Sch. Bip.) Fernald, incl. C. *tuberosa*
   Nutt.)
Arnoglossum plantagineum Raf.   Fen Indian-
   plantain, Tuberous Indian-plantain.
   (*Cacalia plantaginea* (Raf.) Shinners)
† **Arnoseris minima** (L.) Schweigg. & Körte
   Dwarf Nipplewort.
† **Artemisia absinthium** L.   Common
   Wormwood, Absinthe.
Artemisia annua L.   Sweet Wormwood,
   Annual Wormwood.
Artemisia biennis Willd.   Biennial
   Wormwood.
Artemisia campestris L.
   var. canadensis (Michx.) Welsh   Canada
      Wormwood.   (*A. canadensis* Michx.)
   var. caudata (Michx.) E. J. Palmer & Steyerm.
      Beach Wormwood.   (*A. caudata* Michx.)

†**Artemisia gmelinii** Webb. ex Stechm. Summer-fir. (Incl. *A. sacrorum* Ledeb.)

*__Artemisia ludoviciana__ Nutt. Western Mugwort, White Sage. (Incl. var. *gnaphalodes* (Nutt.) Torr. & A. Gray)

†**Artemisia pontica** L. Roman Wormwood.

†**Artemisia stelleriana** Besser Dusty-miller, Beach Wormwood.

*__Artemisia vulgaris__ L. Common Mugwort.

Aster acuminatus Michx. Mountain Aster, Whorled Aster.

Aster borealis (Torr. & A. Gray) Prov. Northern Bog Aster. (Incl. *A. junciformis* Rydb.)

*__Aster brachyactis__ S. F. Blake Western Annual Aster. (*Brachyactis ciliata* (Ledeb.) Ledeb. subsp. *angusta* (Lindl.) A. G. Jones)

Aster cordifolius L. Blue Wood Aster, Common Blue Heart-leaved Aster.

Aster divaricatus L. White Wood Aster, Common White Heart-leaved Aster.

Aster drummondii Lindl. Drummond's Aster, Hairy Heart-leaved Aster.

Aster dumosus L. Bushy Aster, Long-stalked Aster. (Incl. var. *dodgei* Fernald)

Aster ericoides L. White Heath Aster, Squarrose White Aster. (Incl. var. *prostratus* (Kuntze) S. F. Blake)

Aster firmus Nees Shining Aster. (*A. puniceus* var. *firmus* (Nees) Torr. & A. Gray, incl. *A. puniceus* var. *lucidulus* A. Gray)

Aster infirmus Michx. Weak Aster, Appalachian Flat-topped White Aster.

Aster laevis L. Smooth Aster.

Aster lanceolatus Willd. Eastern Lined Aster.
> var. **interior** (Wiegand) Semple & Chmiel. (subsp. *interior* (Wiegand) A. G. Jones, *A. simplex* Willd. var. *interior* (Wiegand) Cronquist)
> var. **lanceolatus** (subsp. *lanceolatus*; incl. *A. simplex* Willd. var. *ramosissimus* (Torr. & A. Gray) Cronquist)
> var. **simplex** (Willd.) A. G. Jones (subsp. *simplex* (Willd.) A. G. Jones, *A. simplex* Willd. var. *simplex*)

Aster lateriflorus (L.) Britton Starved Aster, Goblet Aster, Calico Aster.

Aster linariifolius L. Stiff Aster. (*Ionactis linariifolius* (L.) Greene)

Aster lowrieanus Porter Smooth Heart-leaved Aster.
> var. **lanceolatus** Porter (*A. cordifolius* var. *lanceolatus* Porter)
> var. **lowrieanus** (Incl. *A. cordifolius* var. *laevigatus* Porter)

Aster macrophyllus L. Large-leaved Aster.

Aster novae-angliae L. New England Aster.

Aster oblongifolius Nutt. Aromatic Aster, Shale Barren Aster.

Aster ontarionis Wiegand Bottomland Aster.

Aster oolentangiensis Riddell Prairie Heart-leaved Aster. (Incl. *A. azureus* Lindl.)

Aster patens Aiton Clasping Aster.
> var. **patens**
> var. **phlogifolius** (Muhl. ex Willd.) Nees

Aster paternus Cronquist Toothed White-topped Aster.

Aster pilosus Willd.
> var. **pilosus** Awl Aster.
> var. **pringlei** (A. Gray) S. F. Blake Pringle's Aster. (Incl. var. *demotus* S. F. Blake)

Aster praealtus Poir. Veiny Lined Aster.

Aster prenanthoides Muhl. ex Willd. Crooked-stemmed Aster, Zigzag Aster.

Aster puniceus L. Bristly Aster.

Aster racemosus Elliott Small-headed Aster. (Treated by some authors as *A. vimineus* Lam.)

Aster sagittifolius Wedem. ex Willd. Arrow-leaved Aster. (*A. cordifolius* var. *sagittifolius* (Wedem. ex Willd.) A. G. Jones)

Aster schreberi Nees Glandless Big-leaved Aster.

Aster shortii Lindl. Short's Aster, Midwest Blue Heart-leaved Aster.

Aster solidagineus Michx. Narrow-leaved Aster.

*__Aster subulatus__ Michx. Annual Salt-marsh Aster.

Aster surculosus Michx. Creeping Aster.

†**Aster tataricus** L. f. Tatarian Aster.

†**Aster tradescantii** L. Shore Aster.

(*A. lateriflorus* var. *hirsuticaulis* (Lindl. ex DC.)
Porter)

Aster umbellatus Mill.   TALL FLAT-TOPPED
WHITE ASTER.

Aster undulatus L.   CLASPING HEART-LEAVED
ASTER.

*Bellis perennis L.   ENGLISH DAISY.

Bidens aristosa (Michx.) Britton   MIDWEST
TICKSEED-SUNFLOWER.

Bidens bipinnata L.   SPANISH-NEEDLES.

Bidens cernua L.   NODDING STICK-TIGHT, BUR-
MARIGOLD.

Bidens coronata (L.) Britton   NORTHERN
TICKSEED-SUNFLOWER.

Bidens discoidea (Torr. & A. Gray) Britton
FEW-BRACTED BEGGAR-TICKS.

Bidens frondosa L.   DEVIL'S BEGGAR-TICKS,
DEVIL'S STICK-TIGHT.

Bidens polylepis S. F. Blake   OZARK TICKSEED-
SUNFLOWER.

Bidens tripartita L.   PURPLE-STEMMED BEGGAR-
TICKS.   (Incl. *B. connata* Muhl. ex Willd.)

Bidens vulgata Greene   TALL BEGGAR-TICKS.

Boltonia asteroides (L.) L'Hér. **var. recognita**
(Fernald & Griscom) Cronquist   BOLTONIA,
FALSE ASTER.

†Calendula officinalis L.   CALENDULA.

*Carduus acanthoides L.   PLUMELESS-THISTLE.

*Carduus nutans L.   NODDING-THISTLE, MUSK-
THISTLE.

†Carthamus tinctorius L.   SAFFLOWER, FALSE
SAFFRON.

*Centaurea cyanus L.   CORNFLOWER,
BACHELOR'S-BUTTON.

*Centaurea diffusa Lam.   TUMBLE KNAPWEED,
WHITE-FLOWERED KNAPWEED.

*Centaurea dubia Suter   SHORT-FRINGED
KNAPWEED.   (Treated by some authors as
C. transalpina Schleich. Ex DC.)

*Centaurea jacea L.   BROWN KNAPWEED.

*Centaurea maculosa Lam.   SPOTTED
KNAPWEED.   (Treated by some authors
as *C. biebersteinii* DC.)

*Centaurea nigra L.   BLACK KNAPWEED,
SPANISH-BUTTONS.

*Centaurea repens L.   RUSSIAN KNAPWEED.
(*Acroptilon repens* (L.) DC.)

†Centaurea scabiosa L.   SCABIOSA, HARDHEADS.

*Centaurea solstitialis L.   YELLOW STAR-
THISTLE, BARNABY'S-THISTLE.

*Centaurea × pratensis Thuill. **(Centaurea
jacea × C. nigra)**   INTERMEDIATE KNAPWEED.

†Chrysanthemum balsamita L.   COSTMARY.
(*Balsamita major* Desf.)

†Chrysanthemum coccineum Willd.
PYRETHRUM, PAINTED DAISY.

*Chrysanthemum leucanthemum L.   OX-EYE
DAISY, WHITE DAISY.   (Incl. var. *pinnatifidum*
Lecoq & Lamotte; *Leucanthemum vulgare* Lam.)

†Chrysanthemum maximum Ramond   DAISY
CHRYSANTHEMUM.

*Chrysanthemum parthenium (L.) Bernh.
FEVERFEW.   (*Tanacetum parthenium* (L.)
Sch. Bip.)

†Chrysanthemum segetum L.   CORN-
CHRYSANTHEMUM, CORN-MARIGOLD.

Chrysogonum virginianum L.   GOLDEN-KNEES,
GOLDEN-STAR.

Chrysopsis mariana (L.) Elliott   GOLDEN-ASTER.

*Cichorium intybus L.   CHICORY, BLUE SAILOR.

Cirsium altissimum (L.) Hill   TALL THISTLE.

*Cirsium arvense (L.) Scop.   CANADA THISTLE.
*var. arvense
*var. horridum Wimm. & Grab.

†Cirsium canescens Nutt.   PRAIRIE THISTLE.
(Incl. *C. plattense* (Rydb.) Cockerell ex
Daniels)

Cirsium carolinianum (Walt.) Fernald & B. G.
Schub.   CAROLINA THISTLE, SPRING THISTLE.

Cirsium discolor (Muhl. ex Willd.) Spreng.
FIELD THISTLE.

Cirsium muticum Michx.   SWAMP THISTLE.

Cirsium pumilum (Nutt.) Spreng.   PASTURE
THISTLE.   (Incl. some Ohio specimens previously identified as *C. hillii* (Canby) Fernald)

*Cirsium vulgare (Savi) Ten.   BULL THISTLE,
COMMON THISTLE.

Conyza canadensis (L.) Cronquist   COMMON
HORSEWEED.
var. canadensis
var. pusilla (Nutt.) Cronquist

Conyza ramosissima Cronquist   BUSHY
HORSEWEED, DWARF FLEABANE.

*Coreopsis grandiflora R. Hogg ex Sweet

Large-flowered Tickseed, Large-flowered Coreopsis.

Coreopsis lanceolata L.   Long-stalked Tickseed, Long-stalked Coreopsis.

Coreopsis major Walter   Forest Tickseed, Forest Coreopsis.

*****Coreopsis tinctoria** Nutt.   Golden Tickseed, Plains Tickseed, Golden Coreopsis, Plains Coreopsis.

Coreopsis tripteris L.   Tall Tickseed, Tall Coreopsis.

†**Coreopsis verticillata** L.   Thread-leaved Tickseed, Thread-leaved Coreopsis.

†**Cosmos bipinnatus** Cav.   Cosmos.

*****Crepis biennis** L.   Rough Hawk's-beard.

*****Crepis capillaris** (L.) Wallr.   Smooth Hawk's-beard.

†**Crepis nicaeensis** Balb. ex Pers.   Nicaean Hawk's-beard.

*****Crepis pulchra** L.   Handsome Hawk's-beard.

*****Crepis tectorum** L.   Narrow-leaved Hawk's-beard.

*****Dyssodia papposa** (Vent.) Hitchc.   Fetid-marigold, Stinking-marigold.

Echinacea purpurea (L.) Moench   Purple Coneflower.

Eclipta prostrata (L.) L.   Yerba-de-tajo.   (Incl. *E. alba* (L.) Hassk.)

Elephantopus carolinianus Raeusch.   Elephant's-foot.

Erechtites hieracifolia (L.) Raf. ex DC.   Fireweed, Pilewort.   (Incl. var. *intermedia* Fernald)

Erigeron annuus (L.) Pers.   Daisy Fleabane, White-top.

Erigeron philadelphicus L.   Philadelphia Fleabane, Philadelphia Daisy.

Erigeron pulchellus Michx.
   var. brauniae Fernald   Kentucky Robin's-plantain.
   var. pulchellus   Robin's-plantain.

Erigeron strigosus Muhl. ex Willd.   Rough Fleabane, Rough White-top.

Eupatorium album L.   White Thoroughwort.

Eupatorium altissimum L.   Tall Thoroughwort.

Eupatorium coelestinum L.   Mistflower, Wild Ageratum.

Eupatorium fistulosum Barratt   Common Joe-pye-weed, Hollow-stemmed Joe-pye-weed.

Eupatorium hyssopifolium L. **var. laciniatum** A. Gray   Hyssop Thoroughwort.

Eupatorium incarnatum Walter   Pink Thoroughwort.

Eupatorium maculatum L.   Spotted Joe-pye-weed.

Eupatorium perfoliatum L.   Common Boneset.

Eupatorium purpureum L.   Purple Joe-pye-weed.

Eupatorium rotundifolium L.   Round-leaved Thoroughwort.
   var. ovatum (Bigelow) Torr. (subsp. *ovatum* (Bigelow) J. Montgom. & Fairbrothers)
   var. rotundifolium (subsp. *rotundifolium*)

Eupatorium serotinum Michx.   Late-flowering Thoroughwort.

Eupatorium sessilifolium L.   Upland Boneset.

Euthamia graminifolia (L.) Nutt.   Bushy Goldenrod, Flat-topped Goldenrod.   (*Solidago graminifolia* (L.) Salisb.)
   var. graminifolia (*S. graminifolia* var. *graminifolia*)
   var. nuttallii (Greene) W. Stone (*S. graminifolia* var. *nuttallii* (Greene) Fernald)

Euthamia remota Greene   Great Lakes Goldenrod, Great Lakes Flat-topped Goldenrod.   (Treated by some authors as a part of *E. gymnospermoides* Greene or *Solidago gymnospermoides* (Greene) Fernald)

*****Filago vulgaris** Lam.   Cotton-rose.   (Treated by some authors as *F. germanica* (L.) Huds.)

†**Gaillardia pulchella** Foug.   Rose-ring Blanket-flower.

*****Galinsoga parviflora** Cav.   Lesser Quickweed.

*****Galinsoga quadriradiata** Ruiz & Pav.   Common Quickweed.

Gnaphalium macounii Greene   Clammy Cudweed, Winged Cudweed.   (Treated by some authors as a part of *G. viscosum* Kunth)

Gnaphalium obtusifolium L.   Fragrant Cudweed, Cat's-foot.

Gnaphalium purpureum L.   Purple Cudweed.   (*Gamochaeta purpurea* (L.) Cabrera)

*****Gnaphalium uliginosum** L.   Low Cudweed, Marsh Cudweed.

†**Grindelia lanceolata** Nutt.   Spiny-toothed Gumweed.

†**Grindelia squarrosa** (Pursh) Dunal   Curly-top Gumweed.

†**var. serrulata** (Rydb.) Steyerm.

†**var. squarrosa**

†**Guizotia abyssinica** (L. f.) Cass.   Ramtil.

†**Gutierrezia texana** (DC.) Torr. & A. Gray   Texas Snakeweed.

Hasteola suaveolens (L.) Pojark.   Sweet Indian-plantain, Sweet-smelling Indian-plantain, Hastate-leaved Indian-plantain. (*Cacalia suaveolens* L.; *Synosma suaveolens* (L.) Raf. ex Britton)

*****Helenium amarum** (Raf.) H. Rock   Slender-leaved Sneezeweed, Bitterweed.

Helenium autumnale L.   Common Sneezeweed.

*****Helenium flexuosum** Raf.   Naked Sneezeweed, Southern Sneezeweed.

*****Helianthus angustifolius** L.   Narrow-leaved Sunflower.

*****Helianthus annuus** L.   Common Sunflower.

Helianthus decapetalus L.   Forest Sunflower, Thin-leaved Sunflower.

Helianthus divaricatus L.   Woodland Sunflower, Divaricate Sunflower.

Helianthus giganteus L.   Swamp Sunflower, Giant Sunflower.

Helianthus grosseserratus M. Martens   Sawtooth Sunflower.

Helianthus hirsutus Raf.   Hairy Sunflower. (Incl. var. *stenophyllus* Torr. & A. Gray)

*****Helianthus maximilianii** Schrad.   Maximilian Sunflower.

Helianthus microcephalus Torr. & A. Gray   Small Wood Sunflower, Small-headed Sunflower.

Helianthus mollis Lam.   Ashy Sunflower, Hairy Sunflower.

Helianthus occidentalis Riddell   Few-leaved Sunflower, Naked-stemmed Sunflower, Western Sunflower.

*****Helianthus petiolaris** Nutt.   Prairie Sunflower, Plains Sunflower.

Helianthus strumosus L.   Pale-leaved Sunflower, Rough-leaved Sunflower.

Helianthus tuberosus L.   Jerusalem-artichoke.

Helianthus × brevifolius E. Watson (**Helianthus grosseserratus × H. mollis**)   Shortleaf Sunflower.

Helianthus × cinereus Torr. & A. Gray (**Helianthus mollis × H. occidentalis**)   Gray Sunflower.

Helianthus × doronicoides Lam. (**Helianthus giganetus × H. mollis**)   Lamarck's Sunflower.

Helianthus × glaucus Small (**Helianthus divaricatus × H. microcephalus**)   Glaucous Sunflower.

Helianthus × kellermanii Britton (**Helianthus grosseserratus × H. salicifolius** A. Dietr.)   Kellerman's Sunflower.

Helianthus × laetiflorus Pers. (**Helianthus rigidus** (Cass.) Desf. × **H. tuberosus**)   Showy Sunflower.

Helianthus × luxurians E. Watson (**Helianthus giganteus × H. grosseserratus**)   Watson's Sunflower.

*****Helianthus annuus × H. petiolaris**

Heliopsis helianthoides (L.) Sweet   Smooth Ox-eye, Sunflower-everlasting.

var. helianthoides (subsp. *helianthoides*)

var. occidentalis (T. R. Fisher) Steyerm. (subsp. *occidentalis* T. R. Fisher)

†**Heterotheca camporum** (Greene) Shinners   Golden-aster, Camphor-weed. (*Chrysopsis camporum* Greene)

*****Hieracium aurantiacum** L.   Orange Hawkweed, Devil's-paintbrush.

*****Hieracium caespitosum** Dumort.   Yellow King-devil.

†**Hieracium flagellare** Willd.   Whiplash Hawkweed.

*****Hieracium floribundum** Wimm. & Grab.   Glaucous Hawkweed.

Hieracium gronovii L.   Beaked Hawkweed.

Hieracium kalmii L. **var. fasciculatum** (Pursh) Lepage   Canada Hawkweed. (*H. canadense* Michx. var. *fasciculatum* (Pursh) Fernald)

Hieracium longipilum Torr.   Long-bearded Hawkweed, Long-haired Hawkweed.

Hieracium paniculatum L.   Panicled Hawkweed.

*Hieracium pilosella L.   Mouse-ear
Hawkweed.

*Hieracium piloselloides Vill.   Glaucous
King-devil.   (Incl. *H. florentinum* All.)

Hieracium scabrum Michx.   Rough
Hawkweed, Sticky Hawkweed.

Hieracium venosum L.   Veined Hawkweed,
Rattlesnake-weed.   (Incl. some Ohio specimens previously identified as *H. traillii* Greene
or *H. greenei* A. Gray)

Hieracium gronovii × H. scabrum

Hieracium paniculatum × H. scabrum

Hymenoxys herbacea (Greene) Cusick
Lakeside Daisy.   (*Tetraneuris herbacea*
Greene; treated by some authors as *Hymenoxys
acaulis* (Pursh) K. L. Parker)

*Hypochaeris radicata L.   Spotted Cat's-ear,
Long-rooted Cat's-ear.

*Inula helenium L.   Elecampane.

*Iva annua L.   Rough Marsh-elder,
Sumpweed.

†Iva xanthifolia Nutt.   Big Marsh-elder.

Krigia biflora (Walter) S. F. Blake   Orange
Dwarf-dandelion.

Krigia dandelion (L.) Nutt.   Potato-
dandelion, Colonial Dwarf-dandelion.

Krigia virginica (L.) Willd.   Virginia Dwarf-
dandelion.

Kuhnia eupatorioides L.   False Boneset.
(Incl. var. *corymbulosa* Torr. & A. Gray; *Brick-
ellia eupatorioides* (L.) Shinners)

Lactuca biennis (Moench) Fernald   Biennial
Blue Lettuce, Tall Blue Lettuce.

Lactuca canadensis L.   Wild Lettuce, Tall
Lettuce.
    var. canadensis
    var. latifolia Kuntze
    var. longifolia (Michx.) Farw.
    var. obovata Wiegand

Lactuca floridana (L.) Gaertn.   Woodland
Lettuce, Tall Blue Lettuce.
    var. floridana
    var. villosa (Jacq.) Cronquist

Lactuca hirsuta Muhl. ex Nutt.   Hairy
Lettuce, Hairy Tall Lettuce.

†Lactuca pulchella (Pursh) DC.   Western
Blue Lettuce.   (*L. tatarica* (L.) C. A. Mey.
var. *pulchella* (Pursh) Breitung)

*Lactuca saligna L.   Willow-leaved Lettuce.

*Lactuca sativa L.   Garden Lettuce.

*Lactuca serriola L.   Prickly Lettuce.
    *var. integrata Gren. & Godr.
    *var. serriola

*Lapsana communis L.   Nipplewort.

*Leontodon autumnalis L.   Fall-dandelion.

*Leontodon hispidus L.   Common Hawkbit,
Big Hawkbit.   (Incl. *L. hastilis* L.)

*Leontodon taraxacoides (Vill.) Merat   Little
Hawkbit.   (Incl. *L. leysseri* (Wallr.) Beck)

Liatris aspera Michx.   Lacerate Blazing-star,
Rough Blazing-star.
    var. aspera
    var. intermedia (Lunell) Gaiser

Liatris cylindracea Michx.   Slender Blazing-
star, Few-headed Blazing-star.

†Liatris punctata Hook.   Dotted Blazing-
star.

*Liatris pycnostachya (L.) Willd.   Thick-
spiked Blazing-star, Kansas Gay-feather.

Liatris scariosa (L.) Willd. var. nieuwlandii
(Lunell) E. G. Voss   Large Blazing-star.

Liatris spicata (L.) Willd.   Dense Blazing-
star, Sessile Blazing-star, Spiked Blazing-
star.

Liatris squarrosa (L.) Michx.   Scaly Blazing-
star, Plains Blazing-star.

*Matricaria discoidea DC.   Pineapple-weed.
(*M. matricarioides* (Less.) Porter)

*Matricaria maritima L.   Scentless False
Chamomile.   (Incl. var. *agrestis* (Knaf)
Wilmott)

*Matricaria recutita L.   False Chamomile.
(Treated by some authors as *M. chamomilla* L.)

Megalodonta beckii (Torr. ex Spreng.) Greene
Water-marigold.   (*Bidens beckii* Torr. ex
Spreng.)

Mikania scandens (L.) Willd.   Climbing
Hempweed.

*Onopordum acanthium L.   Scotch-thistle.

†Parthenium hysterophorus L.   Santa Maria.

Parthenium integrifolium L.   Wild Quinine,
Eastern Parthenium.

*Petasites hybridus (L.) P. Gaertn., B. Mey. &
Scherb.   Butterfly-dock, Butterbur.

†Picris echioides L.   Bristly Ox-tongue,
Bristly Picris.

†Picris hieracioides L.   Hawkweed Ox-
tongue.

Pityopsis graminifolia (Michx.) Nutt. **var. latifolia** (Fernald) Semple & F. D. Bowers   Silk-
grass.   (*Chrysopsis graminifolia* (Michx.)
Elliott var. *latifolia* Fernald)

Pluchea camphorata (L.) DC.   Marsh-
fleabane, Camphor-weed.

Polymnia canadensis L.   Leafcup, Pale-
flowered Leafcup.

Polymnia uvedalia (L.) L.   Bear's-foot,
Yellow-flowered Leafcup.   (*Smallanthus
uvedalia* (L.) Mack. ex Small)

Prenanthes alba L.   White Rattlesnake-root,
White-lettuce.

Prenanthes altissima L.   Tall Rattlesnake-
root, Tall White-lettuce.

Prenanthes aspera Michx.   Rough
Rattlesnake-root, Rough White-lettuce.

Prenanthes crepidinea Michx.   Corymbed
Rattlesnake-root, Nodding Rattlesnake-
root, Midwest White-lettuce.

Prenanthes racemosa Michx.   Prairie
Rattlesnake-root, Glaucous White-
lettuce.

Prenanthes serpentaria Pursh   Lion's-foot.

Prenanthes trifoliata (Cass.) Fernald   Gall-of-
the-earth.

†Ratibida columnifera (Nutt.) Wooton &
Standl.   Columnar Coneflower, Mexican-
hat.

Ratibida pinnata (Vent.) Barnhart   Prairie
Coneflower, Gray-headed Coneflower,
Globular Coneflower.

Rudbeckia fulgida Aiton   Orange
Coneflower, Eastern Coneflower.
var. fulgida
var. sullivantii (Boynton & Beadle) Cronquist
var. umbrosa (Boynton & Beadle) Cronquist

Rudbeckia hirta L. **var. pulcherrima** Farw.
Black-eyed Susan.

Rudbeckia laciniata L.
*var. hortensa L. H. Bailey   Golden-glow.
var. laciniata   Cut-leaved Coneflower.

Rudbeckia triloba L.   Three-lobed
Coneflower.

Senecio anonymus A. W. Wood   Appalachian
Squaw-weed, Appalachian Groundsel.

Senecio aureus L.   Golden Ragwort, Heart-
leaved Groundsel.

*Senecio glabellus Poir.   Butterweed, Yellow-
top.

Senecio obovatus Muhl. ex Willd.   Round-
leaved Squaw-weed, Running Groundsel.

Senecio pauperculus Michx.   Balsam Squaw-
weed, Northern Meadow Groundsel.

Senecio plattensis Nutt.   Platte Squaw-weed,
Platte Groundsel.

†Senecio sylvaticus L.   Woodland Squaw-
weed, Woodland Groundsel.

*Senecio vulgaris L.   Common Groundsel.

Silphium laciniatum L.   Compass-plant.

Silphium perfoliatum L.   Cup-plant.

Silphium terebinthinaceum Jacq.
var. lucy-brauniae Steyerm.   Braun's
Prairie-dock.
var. terebinthinaceum   Prairie-dock.

Silphium trifoliatum L.   Whorled Rosinweed.

Silphium laciniatum × S. terebinthinaceum

†Silybum marianum (L.) Gaertn.   Milk-
thistle, St. Mary's-thistle.

Solidago arguta Aiton   Cut-leaved
Goldenrod, Forest Goldenrod.

Solidago bicolor L.   White Goldenrod,
Silver-rod.

Solidago caesia L.   Blue-stemmed
Goldenrod.

Solidago canadensis L.
var. canadensis   Canada Goldenrod
var. rupestris (Raf.) Porter (*S. rupestris* Raf.)
Riverbank Goldenrod.
var. scabra Torr. & A. Gray (*S. altissima* L.)
Tall Goldenrod.

Solidago erecta Pursh   Slender Goldenrod.
(*S. speciosa* var. *erecta* (Pursh) MacMill.)

Solidago flexicaulis L.   Zigzag Goldenrod.

Solidago gigantea Aiton   Smooth Goldenrod.
(Incl. var. *leiophylla* Fernald)

Solidago hispida Muhl. ex Willd.   Hairy
Goldenrod.

Solidago juncea Aiton   Plume Goldenrod,
Early Goldenrod.

Solidago nemoralis Aiton   Gray Goldenrod.

Solidago odora Aiton   Fragrant Goldenrod,
Sweet Goldenrod.

Solidago ohioensis Riddell   Ohio Goldenrod.

Solidago patula Muhl. ex Willd. **var. patula**
ROUGH-LEAVED GOLDENROD.

Solidago ptarmicoides (Nees) B. Boivin WHITE
UPLAND GOLDENROD. (*Aster ptarmicoides*
(Nees) Torr. & A. Gray)

Solidago puberula Nutt. DUSTY GOLDENROD.

Solidago riddellii Frank ex Riddell RIDDELL'S
GOLDENROD.

Solidago rigida L. STIFF GOLDENROD.
　var. glabrata E. L. Braun (*S. jacksonii* (Kuntze)
　　Fernald)
　var. rigida

Solidago rugosa Mill. WRINKLED-LEAF
GOLDENROD.
　var. aspera (Aiton) Fernald
　var. rugosa

*****Solidago sempervirens** L. SEASIDE
GOLDENROD.

Solidago speciosa Nutt. SHOWY GOLDENROD.
　var. rigidiuscula Torr. & A. Gray
　var. speciosa

Solidago sphacelata Raf. SHORT-PAPPUSED
GOLDENROD, FALSE GOLDENROD.

Solidago squarrosa Muhl. STOUT GOLDENROD,
LEAFY GOLDENROD. (Treated by some
authors as a part of *S. petiolaris* Ait.)

Solidago uliginosa Nutt. BOG GOLDENROD.
(Incl. *S. purshii* Porter)

Solidago ulmifolia Muhl. ex Willd. ELM-LEAVED
GOLDENROD.

*****Sonchus arvensis** L. FIELD SOW-THISTLE,
PERENNIAL SOW-THISTLE.
　*****var. arvensis** (subsp. *arvensis*)
　*****var. glabrescens** Günther, Grab. & Wimm.
　　(subsp. *uliginosus* (M. Bieb.) Nyman,
　　S. uliginosus M. Bieb.)

*****Sonchus asper** (L.) Hill SPINY-LEAVED SOW-
THISTLE, PRICKLY SOW-THISTLE.

*****Sonchus oleraceus** L. COMMON SOW-THISTLE.

*****Tanacetum vulgare** L. COMMON TANSY,
GOLDEN-BUTTONS.

*****Taraxacum laevigatum** (Willd.) DC. RED-
SEEDED DANDELION.

*****Taraxacum officinale** Weber ex F. H. Wigg.
COMMON DANDELION.

*****Tragopogon dubius** Scop. FIELD GOAT'S-
BEARD.

*****Tragopogon porrifolius** L. SALSIFY,
VEGETABLE-OYSTER.

*****Tragopogon pratensis** L. YELLOW GOAT'S-
BEARD, SHOWY GOAT'S-BEARD, JACK-GO-TO-
BED-AT-NOON.

*****Tussilago farfara** L. COLTSFOOT.

Verbesina alternifolia (L.) Britton ex Kearney
WINGSTEM.

Verbesina helianthoides Michx. HAIRY
WINGSTEM, OZARK FLATSEED-SUNFLOWER.

Verbesina occidentalis (L.) Walter YELLOW
CROWN-BEARD, SOUTHERN FLATSEED-
SUNFLOWER.

Verbesina virginica L. FROSTWEED, TICKWEED.

Vernonia fasciculata Michx. WESTERN
IRONWEED, PRAIRIE IRONWEED, SMOOTH
IRONWEED.

Vernonia gigantea (Walter) Trel. ex Branner &
Coville TALL IRONWEED.

Vernonia missurica Raf. MISSOURI IRONWEED.

Vernonia noveboracensis (L.) Michx. NEW
YORK IRONWEED.

Vernonia gigantea × **V. noveboracensis**

*****Xanthium spinosum** L. SPINY COCKLEBUR.

Xanthium strumarium L. COMMON
COCKLEBUR.
　var. canadense (Mill.) Torr. & A. Gray
　var. glabratum (DC.) Cronquist

†**Zinnia elegans** Jacq. GARDEN ZINNIA.
(Treated by some authors as a part of
Z. violacea Cav.)

Angiosperms or Flowering Plants: Monocotyledons (Monocots)

Barbara K. Andreas and Tom S. Cooperrider

with the Cyperaceae by Allison W. Cusick and the Orchidaceae by John V. Freudenstein

Phylum **MAGNOLIOPHYTA.**
   ANGIOSPERMS (continued)

Class **Liliopsida.** MONOCOTS

Subclass **ALISMATIDAE**

Order **Alismatales (*Butomales*)**

■ BUTOMACEAE.   FLOWERING-RUSH FAMILY

*****Butomus umbellatus** L.   FLOWERING-RUSH.

■ ALISMATACEAE.   WATER-PLANTAIN FAMILY

Alisma subcordatum Raf.   SOUTHERN WATER-
   PLANTAIN.   (*A. plantago-aquatica* L. var. *parviflorum* (Pursh) Torr.)

Alisma triviale Pursh   NORTHERN WATER-
   PLANTAIN.   (*A. plantago-aquatica* L. var.
   americanum Schult. & Schult. f.)

Echinodorus berteroi (Spreng.) Fassett
   BURHEAD, TALL BURHEAD.   (Incl. var.
   lanceolatus (Engelm. ex S. Watson & J. M.
   Coult.) Fassett and *E. rostratus* (Nutt.)
   Engelm. ex A. Gray)

Sagittaria australis (J. G. Sm.) Small   LONG-
   BEAKED ARROWHEAD, APPALACHIAN ARROW-
   HEAD.

Sagittaria brevirostra Mack. & Bush   MIDWEST
   ARROWHEAD, SHORT-BEAKED ARROWHEAD.

Sagittaria calycina Engelm.   SOUTHERN
   WAPATO.   (*S. montevidensis* Cham. & Schltdl.
   subsp. *calycina* (Engelm.) Bogin; *Lophotocarpus
   calycinus* (Engelm.) J. G. Sm.)

Sagittaria cuneata E. Sheld.   NORTHERN
   ARROWHEAD, WAPATO.

Sagittaria graminea Michx.   GRASS-LEAVED
   ARROWHEAD.

Sagittaria latifolia Willd.   COMMON
   ARROWHEAD, BROAD-LEAVED ARROWHEAD,
   DUCK-POTATO.   (Incl. var. *obtusa* (Engelm.)
   Wiegand and var. *pubescens* (Muhl. ex Nutt.)
   J. G. Sm.)

Sagittaria platyphylla (Engelm.) J. G. Sm.
   ELLIPTIC-LEAVED ARROWHEAD.   (*S. graminea*
   var. *platyphylla* Engelm.)

Sagittaria rigida Pursh   DEER'S-TONGUE
   ARROWHEAD, SESSILE-FRUITED ARROWHEAD.

Order **Hydrocharitales**

■ HYDROCHARITACEAE.   FROG'S-BIT FAMILY

Elodea canadensis Michx.   COMMON
   WATERWEED.

Elodea nuttallii (Planch.) H. St. John   FREE-
   FLOWERED WATERWEED.

Vallisneria americana Michx.   EEL-GRASS,
   TAPE-GRASS, WATER-CELERY.

Order **Najadales**

■ SCHEUCHZERIACEAE.   SCHEUCHZERIA FAMILY

Scheuchzeria palustris L.   SCHEUCHZERIA, POD-
   GRASS.   (Incl. var. *americana* Fernald)

■ JUNCAGINACEAE. Arrow-grass Family

Triglochin maritimum L. Seaside Arrow-grass.

Triglochin palustre L. Marsh Arrow-grass.

■ POTAMOGETONACEAE. Pondweed Family

Potamogeton amplifolius Tuck. Large-leaved Pondweed.

* **Potamogeton crispus** L. Curly-leaved Pondweed, Curly Pondweed.

Potamogeton diversifolius Raf. Common Snailseed Pondweed.

Potamogeton epihydrus Raf. Nuttall's Pondweed, Ribbon-leaved Pondweed. (Incl. var. *nuttallii* (Cham. & Schltdl.) Fernald and var. *ramosus* (Peck) House)

Potamogeton filiformis Pers. **var. alpinus** (Blytt) Asch. & Graebn. Filiform Pondweed. (Incl. var. *borealis* (Raf.) H. St. John; *Stuckenia filiformis* (Pers.) Börner subsp. *alpina* (Blytt) R. R. Haynes, Les, & M. Král)

Potamogeton foliosus Raf. Leafy Pondweed. (Incl. var. *macellus* Fernald)

Potamogeton friesii Rupr. Fries's Pondweed.

Potamogeton gramineus L. Grass-like Pondweed. (Incl. var. *maximus* Morong and var. *myriophyllus* J. W. Robbins)

Potamogeton hillii Morong Hill's Pondweed.

Potamogeton illinoensis Morong Illinois Pondweed.

Potamogeton natans L. Floating Pondweed.

Potamogeton nodosus Poir. Long-leaved Pondweed.

Potamogeton pectinatus L. Fennel-leaved Pondweed, Sago Pondweed. (*Stuckenia pectinata* (L.) Börner)

Potamogeton perfoliatus L. **var. bupleuroides** (Fernald) Farw. Red-headed Pondweed. (subsp. *bupleuroides* (Fernald) Hultén)

Potamogeton praelongus Wulfen White-stemmed Pondweed.

Potamogeton pulcher Tuck. Spotted Pondweed.

Potamogeton pusillus L. Slender Pondweed, Small Pondweed. (Incl. var. *tenuissimus* Mert. & W. D. J. Koch and *P. berchtoldii* Fieber)

Potamogeton richardsonii (A. Benn.) Rydb. Richardson's Pondweed.

Potamogeton robbinsii Oakes Robbins' Pondweed, Fern Pondweed.

Potamogeton spirillus Tuck. Spiral Pondweed, Northern Snailseed Pondweed.

Potamogeton strictifolius A. Benn. Straight-leaved Pondweed.

Potamogeton tennesseensis Fernald Tennessee Pondweed.

Potamogeton vaseyi J. W. Robbins Vasey's Pondweed.

Potamogeton zosteriformis Fernald Flat-stemmed Pondweed.

Potamogeton × hagstroemii A. Benn. (**Potamogeton gramineus × P. richardsonii**) Hagstroem's Pondweed.

Potamogeton × rectifolius A. Benn. (**Potamogeton nodosus × P. richardsonii**) Bennett's Pondweed.

Potamogeton praelongus × P. richardsonii

■ RUPPIACEAE. Ditch-grass Family

† **Ruppia cirrhosa** (Petagna) Grande Ditch-grass, Widgeon-grass. (Until recently, Ohio plants have been treated by most authors as part of *R. maritima* L.)

■ NAJADACEAE. Water-nymph Family

Najas flexilis (Willd.) Rostk. & W. L. E. Schmidt Slender Naiad.

Najas gracillima (A. Braun ex Engelm.) Magnus Thread-like Naiad.

Najas guadalupensis (Spreng.) Magnus Southern Naiad.

***Najas marina** L. Spiny Naiad.

***Najas minor** All. Eurasian Naiad, Eutrophic Naiad, Lesser Naiad.

■ ZANNICHELLIACEAE. Horned-pondweed Family

Zannichellia palustris L. Horned-pondweed.

Subclass **ARECIDAE**

Order **Arales**

■ ACORACEAE. Sweet-flag Family

Acorus americanus (Raf.) Raf. American
  Sweet-flag.
***Acorus calamus** L. Sweet-flag, Calamus.

■ ARACEAE. Arum Family

Arisaema dracontium (L.) Schott Green
  Dragon, Dragon-root.
Arisaema triphyllum (L.) Schott Jack-in-the-
  pulpit. (Incl. var. *pusillum* Peck, var. *stew-
  ardsonii* (Britton) Stevens, A. *stewardsonii*
  Britton, and A. *atrorubens* (Aiton) Blume)
Calla palustris L. Wild Calla.
Peltandra virginica (L.) Schott Tuckahoe,
  Arrow-arum.
Symplocarpus foetidus (L.) Salisb. ex Nutt.
  Skunk-cabbage.

■ LEMNACEAE. Duckweed Family

Lemna minor L. Lesser Duckweed.
Lemna minuta Humb., Bonpl. & Kunth
  Least Duckweed. (Incl. *L. minima* Phil.
  ex Hegelm. and *L. minuscula* Herter; and
  incl. Ohio specimens previously identified
  as *L. valdiviana* Phil.)
Lemna obscura (Austin) Daubs Little
  Duckweed. (*L. minor* var. *obscura* Austin)
Lemna perpusilla Torr. Minute Duckweed.
Lemna trisulca L. Star Duckweed.
Lemna turionifera Landolt Turioned
  Duckweed. (Treated by some authors
  as a part of *L. minor*)
Spirodela polyrhiza (L.) Schleid. Greater
  Duckweed.
Wolffia borealis (Engelm. ex Hegelm.) Landolt
  Dotted Wolffia, Dotted Water-meal.
  (*W. punctata* of authors, not Griseb.)
Wolffia brasiliensis Wedd. Pointed Water-
  meal. (Incl. *W. papulifera* C. H. Thomps.)
Wolffia columbiana H. Karst. American

Wolffia, Common Wolffia, Common
  Water-meal.
Wolffiella gladiata (Hegelm.) Hegelm.
  Wolffiella, Star Wolffiella. (Incl.
  W. floridana (Donn. Sm.) C. H. Thomps.)

Subclass **COMMELINIDAE**

Order **Commelinales**

■ XYRIDACEAE. Yellow-eyed-grass Family

Xyris difformis Chapm. Variable Yellow-
  eyed-grass. (Incl. Ohio plants that have
  until recently been identified as X. *caroliniana*
  Walt.)
Xyris torta Sm. Twisted Yellow-eyed-grass.

■ COMMELINACEAE. Spiderwort Family

***Commelina communis** L. Common
  Dayflower, Asian Dayflower.
***Commelina diffusa** Burm. f. Creeping
  Dayflower.
Commelina virginica L. Virginia Dayflower.
† **Tradescantia bracteata** Small ex Britton
  Sticky Spiderwort, Prairie Spiderwort.
Tradescantia ohiensis Raf. Smooth Spider-
  wort, Ohio Spiderwort.
Tradescantia subaspera Ker Gawl. Zigzag
  Spiderwort.
Tradescantia virginiana L. Virginia
  Spiderwort.

Order **Eriocaulales**

■ ERIOCAULACEAE. Pipewort Family

Eriocaulon aquaticum (Hill) Druce White-
  buttons. (Treated by some authors as
  E. septangulare With.)

Order **Juncales**

■ JUNCACEAE. Rush Family

Juncus acuminatus Michx. Sharp-fruited
  Rush.
Juncus alpinoarticulatus Chaix Alpine Rush.
  (Incl. *J. alpinus* Vill. var. *fuscescens* Fernald and
  var. *rariflorus* Hartm.)

Juncus anthelatus (Wiegand) R. E. Brooks
BRANCHED RUSH. (*J. tenuis* var. *anthelatus* Wiegand)

Juncus articulatus L. JOINTED RUSH.

Juncus balticus Willd. BALTIC RUSH. (*J. articus* Willd. var. *balticus* (Willd.) Trautv., incl. var. *littoralis* Engelm.)

Juncus brachycarpus Engelm. SHORT-FRUITED RUSH.

Juncus brachycephalus (Engelm.) Buchenau SHORT-HEADED RUSH.

Juncus bufonius L. TOAD RUSH.

Juncus canadensis J. Gay ex Laharpe CANADA RUSH.

†**Juncus compressus** Jacq. ROUND-FRUITED RUSH, EURASIAN BLACK-GRASS.

Juncus diffusissimus Buckley DIFFUSE RUSH, SLIM-POD RUSH.

Juncus dudleyi Wiegand DUDLEY'S RUSH. (*J. tenuis* var. *dudleyi* (Wiegand) F. J. Herm.)

Juncus effusus L. SOFT RUSH, COMMON RUSH. (Incl. var. *decipiens* Buchenau, var. *pylaei* (Laharpe) Fernald & Wiegand, and var. *solutus* Fernald & Wiegand)

*__Juncus gerardii__ Loisel. BLACK-GRASS.

Juncus greenei Oakes & Tuck. GREENE'S RUSH.

Juncus interior Wiegand INLAND RUSH, INTERIOR RUSH. (Treated by some authors as a part of *J. tenuis* var. *tenuis*)

Juncus marginatus Rostk. GRASS-LEAVED RUSH. (Incl. *J. biflorus* Elliott)

Juncus nodosus L. KNOTTED RUSH.

Juncus platyphyllus (Wiegand) Fernald FLAT-LEAVED RUSH. (*J. tenuis* var. *platyphyllus* (Wiegand) F. J. Herm.; treated by some authors as a part of *J. dichotomus* Elliott)

Juncus secundus P. Beauv. ex Poir. ONE-SIDED RUSH.

Juncus subcaudatus (Engelm.) Coville & S. F. Blake SHORT-TAILED RUSH.

Juncus tenuis Willd. **var. tenuis** PATH RUSH. (Incl. var. *williamsii* Fernald)

Juncus torreyi Coville TORREY'S RUSH.

Juncus × stuckeyi M. Reinking (**Juncus alpino-articulatus × J. torreyi**) STUCKEY'S RUSH.

Luzula acuminata Raf. HAIRY WOODRUSH.
   var. acuminata EVERGREEN WOODRUSH.

(Incl. *L. carolinae* S. Watson var. *saltuensis* (Fernald) Fernald)
   var. carolinae (S. Watson) Fernald CAROLINA WOODRUSH. (*L. carolinae* S. Watson var. *carolinae*)

Luzula bulbosa (A. W. Wood) Rydb. SOUTHERN WOODRUSH.

Luzula echinata (Small) F. J. Herm. ROUND-LEAVED WOODRUSH. (Incl. var. *mesochorea* F. J. Herm.)

Luzula multiflora (Retz.) Lej. COMMON WOODRUSH, MANY-FLOWERED WOODRUSH, PALE WOODRUSH.

Order **Cyperales**

■ CYPERACEAE. SEDGE FAMILY

by ALLISON W. CUSICK

Bolboschoenus fluviatilis (Torr.) J. Soják RIVER BULRUSH. (*Schoenoplectus fluviatilis* (Torr.) M. T. Strong, *Scirpus fluviatilis* (Torr.) A. Gray)

Bulbostylis capillaris (L.) Kunth ex C. B. Clarke THREAD-LEAVED SEDGE, HAIR SEDGE.

Carex abscondita Mack. SOUTHERN LEAFY WOOD SEDGE.

Carex aggregata Mack. GLOMERATE SEDGE. (*C. sparganioides* var. *aggregata* (Mack.) Gleason)

Carex alata Torr. BROAD-WINGED SEDGE.

Carex albicans Willd. ex Spreng.
   var. albicans OAK SEDGE. (Incl. *C. artitecta* Mack.)
   var. emmonsii (Dewey ex Torr.) Rettig EMMONS' SEDGE. (*C. emmonsii* Dewey ex Torr.)

Carex albolutescens Schwein. PALE STRAW SEDGE.

Carex albursina E. Sheld. WING-STEMMED WOOD SEDGE, WHITE BEAR SEDGE.

Carex alopecoidea Tuck. NORTHERN FOX SEDGE.

Carex amphibola Steud.
   var. amphibola SOUTHERN GRAY WOOD SEDGE.
   var. rigida (L. H. Bailey) Fernald STIFF GRAY WOOD SEDGE.

var. turgida Fernald   GRAY WOOD SEDGE.

Carex annectens (E. P. Bicknell) E. P. Bicknell
YELLOW FOX SEDGE.   (*C. vulpinoidea* var.
ambigua Boott)

Carex appalachica J. M. Webber & P. W. Ball
APPALACHIAN SEDGE.

Carex aquatilis Wahlenb. **var. substricta** Kük.
LEAFY TUSSOCK SEDGE.   (Incl. var. *altior*
(Rybd.) Fernald)

Carex arctata W. Boott ex Hook.   DROOPING
WOOD SEDGE.

Carex argyrantha Tuck.   SILVERY SEDGE.

Carex atherodes Spreng.   WHEAT SEDGE.

Carex atlantica L. H. Bailey
var. atlantica   PRICKLY BOG SEDGE.   (subsp.
atlantica; incl. *C. incomperta* E. P. Bicknell)
var. capillacea (L. H. Bailey) Cronquist
HOWE'S SEDGE.   (subsp. *capillacea* (L. H.
Bailey) Reznicek; *C. howei* Mack.)

Carex aurea Nutt.   GOLDEN-FRUITED SEDGE.

Carex bebbii Olney ex Fernald   BEBB'S SEDGE.

Carex bicknellii Britton   BICKNELL'S SEDGE.

Carex blanda Dewey   COMMON WOOD SEDGE.

Carex brachyglossa Mack.   YELLOW-FRUITED
SEDGE.   (*C. annectens* var. *xanthocarpa* (E. P.
Bicknell) Wiegand)

Carex brevior (Dewey) Mack. ex Lunell
TUFTED FESCUE SEDGE.

Carex bromoides Schkuhr ex Willd.   BROME-
LIKE SEDGE, BROME HUMMOCK SEDGE.

Carex brunnescens (Pers.) Poir. **var. sphaerostachya** (Tuck.) Kük.   BROWNISH SEDGE.
(subsp. *sphaerostachya* (Tuck.) Kalela)

Carex bushii Mack.   BUSH'S SEDGE.

Carex buxbaumii Wahlenb.   BROWN BOG
SEDGE.

Carex canescens L. **var. disjuncta** Fernald
GLAUCOUS SEDGE.   (subsp. *disjuncta* (Fernald)
Toivonen)

Carex careyana Torr. ex Dewey   CAREY'S WOOD
SEDGE.

Carex caroliniana Schwein.   CAROLINA
SEDGE.

Carex cephaloidea (Dewey) Dewey   THIN-
LEAVED SEDGE.   (*C. sparganioides* var. *cephaloidea* (Dewey) J. Carey)

Carex cephalophora Muhl. ex Willd.   OVAL-
HEADED SEDGE.

Carex communis L. H. Bailey   BEECH SEDGE.

Carex comosa Boott   BEARDED SEDGE.

Carex conjuncta Boott   SOFT FOX SEDGE.

Carex conoidea Schkuhr ex Willd.   FIELD
SEDGE.

Carex crawei Dewey   CRAWE'S SEDGE.

Carex crinita Lam.
var. brevicrinis Fernald   SHORT-FRINGED
SEDGE.
var. crinita   TASSELED SEDGE, FRINGED
SEDGE.

Carex cristatella Britton   CRESTED SEDGE.

Carex crus-corvi Shutttlew. ex Kunze   RAVEN-
FOOT SEDGE.

Carex cryptolepis Mack.   LITTLE YELLOW
SEDGE.

Carex davisii Schwein. & Torr.   DAVIS' SEDGE.

Carex debilis Michx.
var. debilis   WEAK SEDGE.
var. rudgei L. H. Bailey   RUDGE'S SEDGE.

Carex decomposita Muhl.   CYPRESS-KNEE
SEDGE.

Carex deweyana Schwein.   DEWEY'S SEDGE.

Carex diandra Schrank   LESSER PANICLED
SEDGE.

Carex digitalis Willd.
var. copulata L. H. Bailey   LARGE NARROW-
LEAVED WOOD SEDGE.   (*C. × copulata*
(L. H. Bailey) Mack.)
var. digitalis   NARROW-LEAVED WOOD
SEDGE.

Carex disperma Dewey   TWO-SEEDED SEDGE.

Carex eburnea Boott   BRISTLE-LEAVED SEDGE.

Carex echinata Murray   LITTLE PRICKLY SEDGE.
(Incl. *C. cephalantha* (L. H. Bailey) E. P.
Bicknell)

Carex emoryi Dewey   EMORY'S SEDGE,
RIVERBANK SEDGE.

Carex festucacea Schkuhr ex Willd.   FESCUE
SEDGE.

Carex flava L.   YELLOW SEDGE.

Carex folliculata L.   LONG-FRUITED SEDGE.

Carex formosa Dewey   HANDSOME SEDGE.

Carex frankii Kunth   FRANK'S SEDGE.

Carex garberi Fernald   GARBER'S SEDGE.

Carex glaucodea Tuck. ex Olney   BLUE-GREEN
SEDGE.   (*C. flaccosperma* Dewey var. *glaucodea*
(Tuck. ex Olney) Kük.)

Carex gracilescens Steud. SLENDER WOOD SEDGE.

Carex gracillima Schwein. GRACEFUL SEDGE.

Carex granularis Muhl. ex Willd. MEADOW SEDGE. (Incl. var. *haleana* (Olney) Porter)

Carex grayi J. Carey GRAY'S SEDGE. (Incl. var. *hispidula* A. Gray)

Carex gynandra Schwein. NODDING SEDGE. (*C. crinita* var. *gynandra* (Schwein.) Torr.)

Carex haydenii Dewey HAYDEN'S SEDGE.

Carex hirsutella Mack. HIRSUTE SEDGE, HAIRY GREEN SEDGE. (*C. complanata* Torr. & Hook. var. *hirsuta* (Willd.) Gleason)

Carex hirtifolia Mack. HAIRY-LEAVED SEDGE, HAIRY WOOD SEDGE.

Carex hitchcockiana Dewey HITCHCOCK'S SEDGE.

Carex hyalinolepis Steud. SWEET MARSH SEDGE.

Carex hystericina Muhl. ex Willd. PORCUPINE SEDGE.

Carex interior L. H. Bailey INLAND SEDGE.

Carex intumescens Rudge BLADDER SEDGE.

Carex jamesii Schwein. JAMES'S SEDGE.

Carex juniperorum Catling, Reznicek & Crins JUNIPER SEDGE.

Carex lacustris Willd. LAKEBANK SEDGE, COMMON LAKE SEDGE.

Carex laevivaginata (Kük.) Mack. SMOOTH-SHEATHED FOX SEDGE.

Carex lasiocarpa Ehrh. **var. americana** Fernald SLENDER SEDGE. (subsp. *americana* (Fernald) Love & Bernard)

Carex laxiculmis Schwein. SPREADING SEDGE, WEAK-STEMMED WOOD SEDGE.

Carex laxiflora Lam. TWO-EDGED SEDGE.
var. laxiflora
var. serrulata F. J. Herm.

Carex leavenworthii Dewey LEAVENWORTH'S SEDGE.

Carex leptalea Wahlenb. BRISTLE-STALKED SEDGE.

Carex leptonervia (Fernald) Fernald NERVELESS SEDGE.

Carex limosa L. MUD SEDGE.

Carex longii Mack. LONG'S SEDGE.

Carex louisianica L. H. Bailey LOUISIANA SEDGE.

Carex lucorum Willd. ex Link FIRE SEDGE.

Carex lupuliformis Sartwell ex Dewey FALSE HOP SEDGE.

Carex lupulina Muhl. ex Willd. HOP SEDGE. (Incl. *C.* × *macounii* Dewey)

Carex lurida Wahlenb. BOTTLEBRUSH SEDGE.

Carex meadii Dewey MEAD'S SEDGE.

Carex merritt-fernaldii Mack. FERNALD'S SEDGE.

Carex mesochorea Mack. MIDLAND SEDGE. (*C. cephalophora* var. *mesochorea* (Mack.) Gleason)

Carex molesta Mack. ex Bright TROUBLESOME SEDGE.

Carex muhlenbergii Schkuhr ex Willd.
var. enervis Boott NERVELESS SAND SEDGE.
var. muhlenbergii MUHLENBERG'S SEDGE.

Carex muskingumensis Schwein. MUSKINGUM SEDGE.

Carex nigromarginata Schwein. BLACK-MARGINED SEDGE.

Carex normalis Mack. LARGER STRAW SEDGE.

Carex oligocarpa Schkuhr ex Willd. FEW-FRUITED SEDGE.

Carex oligosperma Michx. FEW-SEEDED SEDGE.

Carex pallescens L. PALE SEDGE. (Incl. var. *neogaea* Fernald)

Carex peckii Howe PECK'S SEDGE.

Carex pedunculata Muhl. ex Willd. LONG-STALKED SEDGE.

Carex pellita Muhl. WOOLLY SEDGE. (Treated by some authors as *C. lanuginosa* Michx.)

Carex pensylvanica Lam. PENNSYLVANIA SEDGE.

Carex planispicata Naczi FLAT-SPIKED SEDGE.

Carex plantaginea Lam. PLANTAIN SEDGE.

Carex platyphylla J. Carey BROAD-LEAVED WOOD SEDGE.

*Carex praegracilis W. Boott HIGHWAY SEDGE, TOLLWAY SEDGE.

Carex prairea Dewey ex A. W. Wood PRAIRIE SEDGE.

Carex prasina Wahlenb. DROOPING SEDGE.

Carex projecta Mack. NECKLACE SEDGE.

Carex pseudocyperus L.   NORTHERN BEARDED SEDGE.

Carex purpurifera Mack.   PURPLE WOOD SEDGE.

Carex radiata (Wahlenb.) Small   RADIATE SEDGE.   (Treated by some authors as a part of *C. rosea*)

Carex retroflexa Muhl. ex Willd.   REFLEXED SEDGE.

Carex retrorsa Schwein.   REFLEXED BLADDER SEDGE.

Carex richardsonii R. Br.   RICHARDSON'S SEDGE.

Carex rosea Schkuhr ex Willd.   STELLATE SEDGE.   (Incl. *C. convoluta* Mack.)

Carex rugosperma Mack.   LOW SAND SEDGE.

Carex sartwellii Dewey   SARTWELL'S SEDGE.

Carex scabrata Schwein.   SCABROUS SEDGE, ROUGH SEDGE.

Carex scoparia Schkuhr ex Willd.   POINTED BROOM SEDGE.

Carex seorsa Howe   STARRY SEDGE.

Carex shortiana Dewey   SHORT'S SEDGE.

Carex siccata Dewey   HAY SEDGE.   (Treated by some authors as *C. foenea* Willd.)

Carex sparganioides Muhl. ex Willd.   BUR-REED SEDGE.

*****Carex spicata** Huds.   LESSER PRICKLY SEDGE

Carex sprengelii Dewey ex Spreng.   SPRENGEL'S SEDGE.

Carex squarrosa L.   SQUARROSE SEDGE.

Carex sterilis Willd.   FEN SEDGE.

Carex stipata Muhl. ex Willd.
   var. maxima Chapm.   LARGE FOX SEDGE.
   var. stipata   COMMON FOX SEDGE.

Carex straminea Willd. ex Schkuhr   STRAW SEDGE.

Carex striatula Michx.   LINED SEDGE.

Carex stricta Lam.   TUSSOCK SEDGE.

Carex styloflexa Buckley   LOWLAND WOOD SEDGE.

Carex suberecta (Olney) Britton   PRAIRIE STRAW SEDGE.

Carex swanii (Fernald) Mack.   SWAN'S SEDGE.

Carex tenera Dewey   BENDING SEDGE.

Carex tenuiflora Wahlenb.   THIN-FLOWERED SEDGE.

Carex tetanica Schkuhr   STIFF SEDGE.

Carex texensis (G. S. Torr.) L. H. Bailey.   TEXAS SEDGE.   (*C. retroflexa* var. *texensis* (G. S. Torr.) Fernald)

Carex torta Boott ex Tuck.   TWISTED SEDGE.

Carex tribuloides Wahlenb.   BLUNT BROOM SEDGE.

Carex trichocarpa Muhl. ex Willd.   HAIRY-FRUITED SEDGE

Carex trisperma Dewey   THREE-SEEDED SEDGE.

Carex tuckermanii Dewey   TUCKERMAN'S SEDGE.

Carex typhina Michx.   CAT-TAIL SEDGE.

Carex umbellata Schkuhr ex Willd.   CLUSTERED SEDGE.

Carex utriculata Boott   BEAKED SEDGE.   (*C. rostrata* Stokes var. *utriculata* (Boott) L. H. Bailey)

Carex vesicaria L.   INFLATED SEDGE.

Carex virescens Muhl. ex Willd.   GREENISH SEDGE.

Carex viridula Michx.   LITTLE GREEN SEDGE.

Carex vulpinoidea Michx.   FOXTAIL SEDGE.

Carex willdenowii Schkuhr ex Willd.   WILLDENOW'S SEDGE.

Carex woodii Dewey   WOOD'S SEDGE.

Carex × subimpressa Clokey (**Carex hyalinolepis × C. pellita**)   CLOKEY'S LAKE SEDGE.

Carex × sullivantii Boott (**Carex gracillima × C. hirtifolia**)   SULLIVANT'S SEDGE.

Cladium mariscoides (Muhl.) Torr.   TWIG-RUSH.

Cyperus acuminatus Torr. & Hook. ex Torr.   PALE UMBRELLA-SEDGE.

Cyperus bipartitus Torr.   SHINING UMBRELLA-SEDGE.   (Incl. *C. rivularis* Kunth)

Cyperus diandrus Torr.   LOW UMBRELLA-SEDGE.

Cyperus echinatus (L.) A. W. Wood   GLOBOSE UMBRELLA-SEDGE.   (Incl. *C. ovularis* (Michx.) Torr.)

Cyperus erythrorhizos Muhl.   RED-ROOTED UMBRELLA-SEDGE, RED-ROOTED FLAT-SEDGE.

Cyperus esculentus L.   YELLOW NUT-GRASS.

Cyperus flavescens L.   YELLOW UMBRELLA-SEDGE.   (Incl. var. *poiformis* (Pursh) Fernald)

†**Cyperus iria** L.   EURASIAN UMBRELLA-SEDGE.

Cyperus lancastriensis Porter ex A. Gray   MANY-FLOWERED UMBRELLA-SEDGE.

Cyperus lupulinus (Spreng.) Marcks **subsp. macilentus** (Fernald) Marcks SLENDER UMBRELLA-SEDGE. (*C. filiculmis* Vahl var. *macilentus* Fernald)

†**Cyperus microiria** Steud. ASIAN FLAT-SEDGE. (Treated by some authors as *C. amuricus* Maxim.)

Cyperus odoratus L. RUSTY UMBRELLA-SEDGE. (Incl. *C. engelmannii* Steud. and *C. ferruginescens* Boeck.)

Cyperus refractus Engelm. ex Boeck. REFLEXED UMBRELLA-SEDGE.

Cyperus retrofractus (L.) Torr. ROUGH UMBRELLA-SEDGE. (Incl. *C. dipsaciformis* Fernald—TEASEL-SEDGE)

Cyperus schweinitzii Torr. SCHWEINITZ' UMBRELLA-SEDGE.

Cyperus squarrosus L. AWNED UMBRELLA-SEDGE. (Incl. *C. inflexus* Muhl.)

Cyperus strigosus L. GALINGALE.

Cyperus × nieuwlandii Geise (**Cyperus bipartitus × C. flavescens**) NIEUWLAND'S UMBRELLA-SEDGE. (Treated by some authors as a part of *C. flavescens*) .

Dulichium arundinaceum (L.) Britton THREE-WAY SEDGE.

Eleocharis acicularis (L.) Roem. & Schult. NEEDLE SPIKE-RUSH.

Eleocharis compressa Sull. FLAT-STEMMED SPIKE-RUSH.

Eleocharis elliptica Kunth YELLOW-SEEDED SPIKE-RUSH. (*E. tenuis* var. *borealis* (Svenson) Gleason)

Eleocharis engelmannii Steud. ENGELMANN'S SPIKE-RUSH. (*E. ovata* var. *engelmannii* (Steud.) Britton)

Eleocharis erythropoda Steud. RED-FOOTED SPIKE-RUSH. (Incl. *E. calva* Torr.)

Eleocharis geniculata (L.) Roem. & Schult. CARIBBEAN SPIKE-RUSH. (Incl. *E. caribaea* (Rottb.) S. F. Blake)

Eleocharis intermedia Schult. MATTED SPIKE-RUSH.

Eleocharis obtusa (Willd.) Schult. BLUNT SPIKE-RUSH.

Eleocharis olivacea Torr. OLIVACEOUS SPIKE-RUSH. (*E. flavescens* (Poir.) Urb. var. *olivacea* (Torr.) Gleason)

Eleocharis ovata (Roth) Roem. & Schult. OVATE SPIKE-RUSH.

Eleocharis parvula (Roem. & Schult.) Link ex Bluff, Nees & Schauer LEAST SPIKE-RUSH.

Eleocharis pauciflora (Lightf.) Link FEW-FLOWERED SPIKE-RUSH. (Incl. var. *fernaldii* Svenson)

Eleocharis quadrangulata (Michx.) Roem. & Schult. FOUR-ANGLED SPIKE-RUSH. (Incl. var. *crassior* Fernald)

Eleocharis rostellata (Torr.) Torr. ARCHING SPIKE-RUSH.

Eleocharis smallii Britton SMALL'S SPIKE-RUSH.

Eleocharis tenuis (Willd.) Schult.
var. tenuis SLENDER SPIKE-RUSH.
var. verrucosa (Svenson) Svenson WESTERN SLENDER SPIKE-RUSH.

Eleocharis wolfii (A. Gray) A. Gray ex Britton WOLF'S SPIKE-RUSH.

Eriophorum gracile W. D. J. Koch SLENDER COTTON-GRASS.

Eriophorum virginicum L. TAWNY COTTON-GRASS.

Eriophorum viridicarinatum (Engelm.) Fernald GREEN COTTON-GRASS.

Fimbristylis autumnalis (L.) Roem. & Schult. AUTUMN SEDGE.

Kyllinga pumila Michx. ANNUAL GREENHEAD-SEDGE. (Incl. *K. tenuifolia* Steud., *Cyperus tenuifolius* (Steud.) Dandy)

Lipocarpha drummondii (Nees) G. C. Tucker DRUMMOND'S DWARF BULRUSH. (*Hemicarpha micrantha* (Vahl) Pax var. *drummondii* (Nees) Friedl.)

Lipocarpha micrantha (Vahl) G. C. Tucker DWARF BULRUSH. (*Hemicarpha micrantha* (Vahl) Pax var. *micrantha*)

Rhynchospora alba (L.) Vahl WHITE BEAK-RUSH.

Rhynchospora capillacea Torr. FEN BEAK-RUSH.

Rhynchospora capitellata (Michx.) Vahl CLUSTERED BEAK-RUSH.

Rhynchospora globularis (Chapm.) Small **var. recognita** Gale GRASS-LIKE BEAK-RUSH.

Schoenoplectus acutus (Muhl. ex Bigelow) Á. Löve & D. Löve HARD-STEMMED BULRUSH. (*Scirpus acutus* Muhl. ex Bigelow)

Schoenoplectus pungens (Vahl) Palla
CHAIRMAKER'S-RUSH. (*Scirpus pungens*
Vahl; incl. Ohio plants that have in the past
been identified as *Scirpus americanus* Pers.)

Schoenoplectus purshianus (Fernald) M. T.
Strong PURSH'S BULRUSH. (*Scirpus purshianus* Fernald, *Scirpus smithii* A. Gray var.
williamsii (Fernald) Beetle)

†**Schoenoplectus saximontanus** (Fernald)
J. Raynal ROCKY MOUNTAIN BULRUSH.
(*Scirpus saximontanus* Fernald)

Schoenoplectus smithii (A. Gray) Soják
SMITH'S BULRUSH. (*Scirpus smithii*
A. Gray var. *smithii*)

Schoenoplectus subterminalis (Torr.) Soják
SWAYING-RUSH. (*Scirpus subterminalis* Torr.)

Schoenoplectus tabernaemontani (C. C. Gmel.)
Palla GREAT BULRUSH, SOFT-STEMMED
BULRUSH. (*Scirpus tabernaemontani* C. C.
Gmel.; incl. *Scirpus validus* Vahl var. *creber*
Fernald and var. *validus*)

Schoenoplectus torreyi (Olney) Palla TORREY'S
BULRUSH. (*Scirpus torreyi* Olney)

Scirpus atrovirens Willd. DARK GREEN
BULRUSH.

Scirpus cyperinus (L.) Kunth WOOL-GRASS.
(Incl. var. *pelius* Fernald and *S. rubricosus*
Fernald)

Scirpus expansus Fernald WOODLAND
BULRUSH.

Scirpus georgianus R. M. Harper GEORGIAN
DARK GREEN BULRUSH. (*S. atrovirens* var.
georgianus (R. M. Harper) Fernald)

Scirpus hattorianus Makino SMOOTH-LEAVED
DARK GREEN BULRUSH.

Scirpus pedicellatus Fernald PEDICELLED
WOOL-GRASS.

Scirpus pendulus Muhl. DROOPING BULRUSH.
(Treated by some authors as a part of *S. line-
atus* Michx.)

Scirpus polyphyllus Vahl LEAFY BULRUSH.

Scleria oligantha Michx. TUBERCLED NUT-
RUSH.

Scleria pauciflora Muhl. ex Willd. FEW-
FLOWERED NUT-RUSH.
  var. **caroliniana** (Willd.) A. W. Wood
  var. **pauciflora**

Scleria triglomerata Michx. TALL NUT-RUSH.

Scleria verticillata Muhl. ex Willd. LOW NUT-
RUSH.

Trichophorum planifolium (Spreng.) Palla
FLAT-LEAVED BULRUSH. (Incl. *Scirpus vere-
cundus* Fernald)

■ POACEAE or GRAMINEAE. GRASS
FAMILY

*****Aegilops cylindrica** Host JOINTED GOAT
GRASS.

†**Agropyron desertorum** (Fisch. ex Link) Schult.
STANDARD CRESTED WHEAT GRASS, DESERT
WHEAT GRASS, RUSSIAN WHEAT GRASS.

*****Agrostis capillaris** L. RHODE ISLAND BENT
GRASS, COLONIAL BENT GRASS. (Incl.
A. tenuis Sibth.)

Agrostis elliottiana Schult. ELLIOTT'S BENT
GRASS.

*****Agrostis gigantea** Roth REDTOP, BLACK BENT
GRASS. (Incl. Ohio plants treated by some
authors as *A. alba* L.)

Agrostis hyemalis (Walter) Britton, Sterns, &
Poggenb. TICKLE GRASS.

Agrostis perennans (Walter) Tuck. AUTUMN
BENT GRASS, UPLAND BENT GRASS.

Agrostis scabra Willd. FLY-AWAY GRASS.
(*A. hyemalis* var. *scabra* (Willd.) Blomq., incl.
A. hyemalis var. *tenuis* (Tuck.) Gleason)

*****Agrostis stolonifera** L. **var. palustris** (Huds.)
Farw. CREEPING BENT GRASS.

†**Aira caryophyllea** L. SILVER HAIR GRASS.

Alopecurus aequalis Sobol. SHORT-AWNED
FOXTAIL.

Alopecurus carolinianus Walter CAROLINA
FOXTAIL, ANNUAL FOXTAIL.

†**Alopecurus myosuroides** Huds. SLENDER
FOXTAIL.

*****Alopecurus pratensis** L. MEADOW FOXTAIL.

Ammophila breviligulata Fernald BEACH
GRASS, AMERICAN BEACH GRASS.

Andropogon gerardii Vitman BIG BLUESTEM,
TURKEY-FOOT GRASS.

Andropogon glomeratus (Walter) Britton,
Sterns, & Poggenb. BUSHY BROOM-SEDGE,
BUSHY BEARD GRASS. (Incl. *A. virginicus* var.
abbreviatus (Hack.) Fernald & Griscom)

Andropogon gyrans Ashe ELLIOTT'S BEARD GRASS. (Incl. A. *elliottii* Chapm.)

Andropogon virginicus L. COMMON BROOM-SEDGE.

†Anthoxanthum aristatum Boiss. ANNUAL SWEET VERNAL GRASS, PUEL'S GRASS.

*Anthoxanthum odoratum L. SWEET VERNAL GRASS.

†Apera interrupta (L.) P. Beauv. INTERRUPTED LITTLE-AWNED GRASS.

†Apera spica-venti (L.) P. Beauv. OPEN LITTLE-AWNED GRASS. (*Agrostis spica-venti* L.)

Aristida dichotoma Michx. POVERTY GRASS, CHURCHMOUSE THREE-AWNED GRASS.

Aristida longespica Poir. var. geniculata (Raf.) Fernald FALSE ARROW-FEATHER, SLIM-SPIKE THREE-AWNED GRASS. (Incl. A. *necopina* Shinners)

Aristida oligantha Michx. PRAIRIE THREE-AWNED GRASS, PLAINS THREE-AWNED GRASS.

Aristida purpurascens Poir. PURPLE THREE-AWNED GRASS, PURPLE TRIPLE-AWNED GRASS.

*Arrhenatherum elatius (L.) P. Beauv. ex J. Presl & C. Presl TALL OAT GRASS.

*Arthraxon hispidus (Thunb.) Makino HAIRY JOINT GRASS.

Arundinaria gigantea (Walter) Muhl. GIANT CANE.

†Avena fatua L. WILD OATS.

†Avena sativa L. OATS, COMMON OATS.

†Beckmannia syzigachne (Steud.) Fernald AMERICAN SLOUGH GRASS.

Bouteloua curtipendula (Michx.) Torr. SIDE-OATS GRAMA GRASS.

†Bouteloua gracilis (Willd. ex Kunth) Lag. ex Griffiths BLUE GRAMA GRASS.

†Bouteloua hirsuta Lag. HAIRY GRAMA GRASS.

Brachyelytrum erectum (Schreb. ex Spreng.) P. Beauv. LONG-AWNED WOOD GRASS.

*Bromus arvensis L. FIELD BROME.

†Bromus briziformis Fisch. & C. A. Mey. RATTLESNAKE CHESS, QUAKE GRASS.

Bromus ciliatus L. FRINGED BROME.

*Bromus commutatus Schrad. HAIRY CHESS.

†Bromus erectus Huds. TALL BROME, ERECT BROME.

†Bromus hordeaceus L. SOFT CHESS. (Treated by some authors as *B. mollis* L.)

*Bromus inermis Leyss. SMOOTH BROME, HUNGARIAN BROME.

*Bromus japonicus Thunb. ex Murray JAPANESE BROME, JAPANESE CHESS.

Bromus kalmii A. Gray PRAIRIE BROME, PLAINS BROME, KALM'S CHESS.

Bromus latiglumis (Shear) Hitchc. EAR-LEAVED BROME. (Treated by some authors as *B. altissimus* Pursh)

Bromus nottawayanus Fernald SATIN BROME.

Bromus pubescens Muhl. ex Willd. CANADA BROME, WOODLAND BROME. (Incl. plants formerly called by the now rejected name *B. purgans* L.)

*Bromus secalinus L. CHEAT, CHESS.

*Bromus sterilis L. BARREN BROME.

*Bromus tectorum L. DOWNY BROME, DOWNY CHESS.

Calamagrostis canadensis (Michx.) P. Beauv. CANADA BLUEJOINT.

†Calamagrostis cinnoides (Muhl.) Barton EASTERN BLUEJOINT. (Treated by some authors as *C. coarctata* (Torr.) D. C. Eaton)

Calamagrostis porteri A. Gray subsp. insperata (Swallen) C. W. Greene BARTLEY'S REED GRASS. (*C. insperata* Swallen)

Calamagrostis stricta (Timm) Koeler NORTHERN REED GRASS, NORTHERN BLUEJOINT. (Incl. subsp. *inexpansa* (A. Gray) C. W. Greene, *C. inexpansa* A. Gray)

*Calamovilfa longifolia (Hook.) Scribn. var. magna Scribn. & Merr. GREATER PRAIRIE SANDREED.

Cenchrus longispinus (Hack.) Fernald COMMON SANDBUR. (Treated by some authors as *C. pauciflorus* Benth.)

Chasmanthium latifolium (Michx.) H. O. Yates BROAD-LEAVED SPIKE GRASS, WILD OATS. (*Uniola latifolia* Michx.)

Cinna arundinacea L. COMMON WOOD-REED.

Cinna latifolia (Trevir. ex R. Goepp.) Griseb. NORTHERN WOOD-REED.

†Crypsis schoenoides (L.) Lam. FALSE TIMOTHY, RAT-TAIL GRASS. (*Heleochloa schoenoides* (L.) Host ex Roem.)

*Cynodon dactylon (L.) Pers.   BERMUDA
   GRASS.
†Cynosurus cristatus L.   CRESTED DOG'S-TAIL
   GRASS.
†Cynosurus echinatus L.   ANNUAL DOG'S-TAIL
   GRASS.
*Dactylis glomerata L.   ORCHARD GRASS.
†Dactyloctenium aegypticum (L.) Willd.
   CROWFOOT GRASS.
Danthonia compressa Austin ex Peck
   FLATTENED WILD OAT GRASS.
Danthonia spicata (L.) P. Beauv. ex Roem. &
   Schult.   POVERTY OAT GRASS.
Deschampsia cespitosa (L.) P. Beauv.   TUFTED
   HAIR GRASS.
†Deschampsia danthonioides (Trin.) Munro
   ANNUAL HAIR GRASS.
Deschampsia flexuosa (L.) Trin.   CRINKLED
   HAIR GRASS, WAVY HAIR GRASS.
Diarrhena americana P. Beauv.   BEAK GRASS,
   AMERICAN BEAK GRASS.
Diarrhena obovata (Gleason) Brandenburg
   WESTERN BEAK GRASS.   (D. americana var.
   obovata Gleason)
*Digitaria ciliaris (Retz.) Koeler   SOUTHERN
   CRAB GRASS.
Digitaria cognata (Schult.) Pilg.   FALL WITCH
   GRASS.   (Leptoloma cognatum (Schult.)
   Chase)
Digitaria filiformis (L.) Koeler   SLENDER CRAB
   GRASS, SLENDER FINGER GRASS.
*Digitaria ischaemum (Schreb.) Muhl.
   SMOOTH CRAB GRASS, SMALL CRAB GRASS.
*Digitaria sanguinalis (L.) Scop.   NORTHERN
   CRAB GRASS, HAIRY CRAB GRASS.
†Distichlis spicata (L.) Greene   SEASHORE
   SALT GRASS.
*Echinochloa crusgalli (L.) P. Beauv.
   BARNYARD GRASS.
†Echinochloa frumentacea Link   JAPANESE
   MILLET, BILLION-DOLLAR GRASS.   (E. crusgalli
   var. frumentacea (Link) W. Wight)
Echinochloa muricata (P. Beauv.) Fernald
   *var. microstachya Wiegand   WIEGAND'S
      BARNYARD GRASS, WESTERN BARNYARD
      GRASS.   (Incl. E. wiegandii (Fassett)
      McNeil & Dore, E. pungens (Poir.) Rydb.
      var. wiegandii Fassett)

   var. muricata   SOUTHERN BARNYARD GRASS.
      (Incl. E. pungens (Poir.) Rydb. var. pungens)
Echinochloa walteri (Pursh) A. Heller
   WALTER'S BARNYARD GRASS, SALT-MARSH
   COCKSPUR GRASS.
*Eleusine indica (L.) Gaertn.   GOOSE GRASS,
   YARD GRASS.
Elymus canadensis L.   CANADA WILD RYE.
Elymus hystrix L.   BOTTLEBRUSH GRASS.
   (Hystrix patula Moench)
*Elymus repens (L.) Gould   QUACK GRASS,
   WHEAT GRASS.   (Agropyron repens (L.)
   P. Beauv., Elytrigia repens (L.) Desv. ex
   B. D. Jacks.)
Elymus riparius Wiegand   STREAMBANK WILD
   RYE, RIVERBANK WILD RYE.
Elymus trachycaulus (Link) Gould ex Shinners
   BEARDED WHEAT GRASS, SLENDER WHEAT
   GRASS.   (Agropyron trachycaulum (Link)
   Malte ex H. F. Lewis, incl. var. glaucum
   (Pease & A. H. Moore) Malte)
Elymus villosus Muhl. ex Willd.   DOWNY WILD
   RYE.
Elymus virginicus L.   VIRGINIA WILD RYE.
   (Incl. var. glabriflorus (Vasey) Bush and var.
   submuticus Hook.)
Eragrostis capillaris (L.) Nees   CAPILLARY LOVE
   GRASS, LACE GRASS.
*Eragrostis cilianensis (All.) Vignolo ex Janch.
   STINK GRASS, SKUNK GRASS.   (Incl. E. mega-
   stachya (Koeler) Link)
†Eragrostis curvula (Schrad.) Nees   WEEPING
   LOVE GRASS.
Eragrostis frankii C. A. Mey. ex Steud.
   SANDBAR LOVE GRASS.
†Eragrostis hirsuta (Michx.) Nees   BIGTOP
   LOVE GRASS.
Eragrostis hypnoides (Lam.) Britton, Sterns, &
   Poggenb.   MOSS-LIKE LOVE GRASS, CREEPING
   LOVE GRASS.
*Eragrostis minor Host   LOW LOVE GRASS.
   (Incl. E. poaeoides P. Beauv. ex Roem. &
   Schult.)
Eragrostis pectinacea (Michx.) Nees ex Steud.
   CAROLINA LOVE GRASS, SMALL LOVE
   GRASS.
*Eragrostis pilosa (L.) P. Beauv.   INDIA LOVE
   GRASS.

Eragrostis spectabilis (Pursh) Steud.   PURPLE LOVE GRASS, SHOWY LOVE GRASS.

†**Eragrostis trichodes** (Nutt.) A. W. Wood SAND LOVE GRASS.

****Festuca ovina** L.   SHEEP FESCUE.   (Incl. Ohio plants treated by some authors as var. *capillata* (Lam.) Alef., and incl. *F. filiformis* Pourr. and *F. tenuifolia* Sibth.)

****Festuca rubra** L.   RED FESCUE.

Festuca subverticillata (Pers.) E. B. Alexeev NODDING FESCUE.   (Incl. *F. obtusa* Biehler)

†**Festuca trachyphylla** (Hack.) Krajina   HARD FESCUE.

Glyceria acutiflora Torr.   SHARP-GLUMED MANNA GRASS.

Glyceria borealis (Nash) Batch.   NORTHERN MANNA GRASS.

Glyceria canadensis (Michx.) Trin. RATTLESNAKE MANNA GRASS.

Glyceria grandis S. Watson   TALL MANNA GRASS.

Glyceria melicaria (Michx.) F. T. Hubb.   LONG MANNA GRASS.

Glyceria septentrionalis Hitchc.   FLOATING MANNA GRASS.

Glyceria striata (Lam.) Hitchc.   FOWL MANNA GRASS.

Gymnopogon ambiguus (Michx.) Britton, Sterns, & Poggenb.   BEARD GRASS.

Hesperostipa spartea (Trin.) Barkworth PORCUPINE GRASS.   (*Stipa spartea* Trin.)

Hierochloe odorata (L.) P. Beauv.   VANILLA GRASS, SWEET GRASS.

****Holcus lanatus** L.   VELVET GRASS.

†**Hordeum geniculatum** All.   MEDITERRANEAN BARLEY, KNEE BARLEY.   (Treated by some authors as a part of *H. marinum* Huds.)

****Hordeum jubatum** L.   SQUIRREL-TAIL BARLEY, FOXTAIL BARLEY.

****Hordeum pusillum** Nutt.   LITTLE BARLEY.

†**Hordeum vulgare** L.   BARLEY.

Koeleria macrantha (Ledeb.) Schult.   JUNE GRASS.   (Treated by some authors as *K. cristata* Pers. or as *K. pyramidata* (Lam.) P. Beauv.)

Leersia lenticularis Michx.   CATCHFLY GRASS.

Leersia oryzoides (L.) Sw.   RICE CUT GRASS.

Leersia virginica Willd.   WHITE GRASS, WOODLAND CUT GRASS.

****Leptochloa fusca** (L.) Kunth   SPRANGLETOP. (Incl. *L. fascicularis* (Lam.) A. Gray var. *fascicularis* and var. *acuminata* (Nash) Gleason, *Diplachne acuminata* Nash, *L. acuminata* (Nash) Mohlenbr.)

****Leptochloa panicea** (Retz.) Ohwi   RED SPRANGLETOP.   (Incl. *L. filiformis* (Lam.) P. Beauv. and *L. mucronata* (Michx.) Kunth)

****Lolium arundinaceum** (Schreb.) Darbysh. TALL FESCUE, ALTA FESCUE.   (*Festuca arundinacea* Schreb., *F. elatior* L. var. *arundinacea* (Schreb.) Wimm., *Schedonorus arundinaceus* (Schreb.) Dumort.)

****Lolium multiflorum** Lam.   ITALIAN RYE GRASS.   (*L. perenne* var. *multiflorum* (Lam.) Parn.)

****Lolium perenne** L.   PERENNIAL RYE GRASS. (Incl. var. *aristatum* Willd.)

****Lolium pratense** (Huds.) Darbysh.   MEADOW FESCUE.   (*Festuca pratensis* Huds., incl. plants formerly called by the now rejected name *F. elatior* L., *Schedonorus pratensis* (Huds.) P. Beauv.)

†**Lolium temulentum** L.   DARNEL.

Melica nitens (Scribn.) Nutt. ex Piper   THREE-FLOWERED MELIC.

****Microstegium vimineum** (Trin.) A. Camus RECLINING EULALIA.   (*Eulalia viminea* (Trin.) Kuntze)

Milium effusum L.   WOOD MILLET.

****Miscanthus sinensis** Andersson   EULALIA.

****Muhlenbergia asperifolia** (Nees & Meyen ex Trin.) Parodi   SCRATCH GRASS.

Muhlenbergia capillaris (Lam.) Trin.   HAIR GRASS.

Muhlenbergia cuspidata (Torr. ex Hook.) Rydb. PLAINS MUHLENBERGIA.

Muhlenbergia frondosa (Poir.) Fernald COMMON SATIN GRASS.

Muhlenbergia glomerata (Willd.) Trin.   MARSH WILD TIMOTHY.

Muhlenbergia mexicana (L.) Trin.   LEAFY SATIN GRASS.

Muhlenbergia schreberi J. F. Gmel. NIMBLEWILL.

Muhlenbergia sobolifera (Muhl. ex Willd.) Trin. ROCK SATIN GRASS.

Muhlenbergia sylvatica (Torr.) Torr. ex A. Gray FOREST SATIN GRASS, WOODLAND MUHLEN-BERGIA.

Muhlenbergia tenuiflora (Willd.) Britton, Sterns, & Poggenb. SLENDER SATIN GRASS.

Muhlenbergia × curtisetosa (Scribn.) Bush (**Muhlenbergia frondosa × M. schreberi**) SHORT-BRISTLED SATIN GRASS.

Oryzopsis asperifolia Michx. LARGE-LEAVED MOUNTAIN-RICE.

Panicum acuminatum Sw. (Treated by some authors as *P. lanuginosum* Elliott; *Dichanthelium acuminatum* (Sw.) Gould & C. A. Clark)

  var. fasiculatum (Torr.) Lelong OLD-FIELD PANIC GRASS. (*P. lanuginosum* var. *fasciculatum* (Torr.) Fernald, incl. var. *implicatum* (Scribn.) Fernald, *P. implicatum* Scribn.; *D. acuminatum* var. *fasciculatum* (Torr.) Freckmann)

  var. lindheimeri (Nash) Lelong LINDHEIMER'S PANIC GRASS. (*P. lanuginosum* var. *lindheimeri* (Nash) Fernald; *D. acuminatum* var. *lindheimeri* (Nash) Gould & C. A. Clark)

Panicum anceps Michx. BEAKED PANIC GRASS.

Panicum bicknellii Nash BICKNELL'S PANIC GRASS. (Treated by some authors as a part of *P. boreale* or *Dichanthelium boreale* (Nash) Freckmann)

Panicum boreale Nash NORTHERN PANIC GRASS. (*Dichanthelium boreale* (Nash) Freckmann)

Panicum boscii Poir. BOSC'S PANIC GRASS. (*Dichanthelium boscii* (Poir.) Gould & C. A. Clark)

Panicum calliphyllum Ashe TALL GREEN PANIC GRASS. (Treated by some authors as a part of *P. boreale* or *Dichanthelium boreale* (Nash) Freckmann)

Panicum capillare L. WITCH GRASS. (Incl. var. *occidentale* Rydb.)

Panicum clandestinum L. DEER'S-TONGUE GRASS. (*Dichanthelium clandestinum* (L.) Gould)

Panicum columbianum Scribn. AMERICAN PANIC GRASS. (Incl. var. *thinium* Hitchc. & Chase, *Dichanthelium sabulorum* (Lam.) Gould & C. A. Clark var. *thinium* (Hitchc. & Chase) Gould & C. A. Clark)

Panicum commonsianum Ashe COMMONS' PANIC GRASS. (Treated by some authors as a part of *P. ovale* Elliott var. *addisonii* (Nash) C. F. Reed, *Dichanthelium ovale* (Elliott) Gould & C. A. Clark var. *addisonii* (Nash) Gould & C. A. Clark)

Panicum commutatum Schult. VARIABLE PANIC GRASS, OVAL-LEAVED PANIC GRASS. (Incl. var. *ashei* (G. Pearson ex Ashe) Fernald; *Dichanthelium commutatum* (Schult.) Gould)

Panicum depauperatum Muhl. STARVED PANIC GRASS. (*Dichanthelium depauperatum* (Muhl.) Gould)

Panicum dichotomiflorum Michx. FALL PANIC GRASS, SPREADING PANIC GRASS, KNEE GRASS.

Panicum dichotomum L. FORKED PANIC GRASS. (Incl. var. *barbulatum* (Michx.) A. W. Wood; *Dichanthelium dichotomum* (L.) Gould)

Panicum flexile (Gatt.) Scribn. WIRY WITCH GRASS.

Panicum gattingeri Nash GATTINGER'S WITCH GRASS. (*P. capillare* var. *campestre* Gatt.)

Panicum latifolium L. BROAD-LEAVED PANIC GRASS. (*Dichanthelium latifolium* (L.) Gould & C. A. Clark)

Panicum laxiflorum Lam. PALE GREEN PANIC GRASS. (*Dichanthelium laxiflorum* (Lam.) Gould)

Panicum leibergii (Vasey) Scribn. LEIBERG'S PANIC GRASS. (*Dichanthelium leibergii* (Vasey) Freckmann)

Panicum linearifolium Scribn. ex Nash SLENDER-LEAVED PANIC GRASS. (Incl. var. *werneri* Scribn.; *Dichanthelium linearifolium* (Scribn. ex Nash) Gould)

Panicum longifolium Torr. LONG-LEAVED PANIC GRASS. (*P. rigidulum* Bosc ex Nees var. *pubescens* (Vasey) Lelong)

Panicum meridionale Ashe SOUTHERN HAIRY PANIC GRASS. (*Dichanthelium meridionale* (Ashe) Freckmann; treated by some authors as a part of *P. leucothrix* Nash)

Panicum microcarpon Muhl. ex Elliott SMALL-FRUITED PANIC GRASS. (Treated by some authors as a part of *P. dichotomum* or *Dichanthelium dichotomum* (L.) Gould)

† Panicum miliaceum L. BROOM-CORN MILLET, PROSO MILLET.

Panicum oligosanthes Schult. **var. scribnerianum** (Nash) Fernald SAND PANIC GRASS, PRAIRIE PANIC GRASS. (*Dichanthelium oligosanthes* (Schult.) Gould var. *scribnerianum* (Nash) Gould)

Panicum perlongum Nash LONG-PANICLED PANIC GRASS. (Treated by some authors as a part of *P. linearifolium* or *Dichanthelium linearifolium* (Scribn. ex Nash) Gould)

Panicum philadelphicum Bernh. ex Trin. PHILADELPHIA PANIC GRASS.

Panicum polyanthes Schult. MANY-FLOWERED PANIC GRASS. (*Dichanthelium sphaerocarpon* (Elliott) Gould var. *isophyllum* (Scribn.) Gould & C. A. Clark)

Panicum praecocius Hitchc. & Chase EARLY PANIC GRASS. (Treated by some authors as a part of *P. villosissimum* Nash, *Dichanthelium villosissimum* (Nash) Freckmann var. *praecocius* (Hitchc. & Chase) Freckmann)

Panicum rigidulum Bosc ex Nees MARSH PANIC GRASS. (Incl. *P. agrostoides* Michx. and *P. stipitatum* Nash)

Panicum sphaerocarpon Elliott ROUND-FRUITED PANIC GRASS. (*Dichanthelium sphaerocarpon* (Elliott) Gould)

Panicum spretum Schult. NARROW-HEADED PANIC GRASS. (*Dichanthelium spretum* (Schult.) Freckmann)

Panicum tuckermanii Fernald TUCKERMAN'S PANIC GRASS. (*P. philadelphicum* var. *tuckermanii* (Fernald) Steyerm. & Schmoll)

Panicum verrucosum Muhl. WARTY PANIC GRASS.

Panicum virgatum L. SWITCH GRASS.

Panicum yadkinense Ashe SPOTTED PANIC GRASS. (Treated by some authors as a part of *P. dichotomum* or *Dichanthelium dichotomum* (L.) Gould)

*Pascopyrum smithii (Rydb.) Á. Löve WESTERN WHEAT GRASS. (*Agropyron smithii* Rydb., *Elytrigia smithii* (Rydb.) Nevski)

† Paspalum floridanum Michx. BEAD GRASS.

Paspalum fluitans (Elliott) Kunth RIVERBANK PASPALUM.

Paspalum laeve Michx. SMOOTH LENS GRASS.

Paspalum pubiflorum Rupr. ex E. Fourn. **var. glabrum** Vasey ex Scribn. RECLINING MEAL GRASS, RECLINING BEAD GRASS.

Paspalum setaceum Michx. **var. ciliatifolium** (Michx.) Vasey SLENDER PASPALUM, SLENDER BEAD GRASS. (Incl. var. *longipedunculatum* Leconte; *P. ciliatifolium* Michx.)

Phalaris arundinacea L. REED CANARY GRASS.

† Phalaris canariensis L. CANARY GRASS.

† Phalaris caroliniana Walter COASTAL CANARY GRASS, MAY GRASS.

*Phleum pratense L. TIMOTHY.

Phragmites australis (Cav.) Trin. ex Steud. REED GRASS, COMMON REED. (Incl. *P. communis* Trin.)

Piptatherum racemosum (Sm.) Eaton MOUNTAIN-RICE, BLACK-SEEDED GRASS. (*Oryzopsis racemosa* (Sm.) Ricker ex Hitchc.)

Piptochaetium avenaceum (L.) Parodi BLACK-SEEDED NEEDLE GRASS, BLACK OAT GRASS. (*Stipa avenacea* L.)

Poa alsodes A. Gray WOOD SPEAR GRASS.

*Poa annua L. ANNUAL BLUE GRASS.

*Poa arida Vasey PLAINS BLUE GRASS.

*Poa bulbosa L. BULBOUS BLUE GRASS.

*Poa chapmaniana Scribn. CHAPMAN'S BLUE GRASS.

*Poa compressa L. CANADA BLUE GRASS.

Poa cuspidata Nutt. CUSPIDATE SPEAR GRASS.

Poa languida Hitchc. WEAK SPEAR GRASS. (Treated by some authors as a part of *P. saltuensis*)

*Poa nemoralis L. WOOD BLUE GRASS.

Poa paludigena Fernald & Wiegand MARSH SPEAR GRASS.

Poa palustris L. FOWL MEADOW GRASS.

*Poa pratensis L. KENTUCKY BLUE GRASS.

Poa saltuensis Fernald & Wiegand PASTURE BLUE GRASS.

Poa sylvestris A. Gray FOREST BLUE GRASS.

*Poa trivialis L. ROUGH BLUE GRASS.

Poa wolfii Scribn. WOLF'S BLUE GRASS.

*Puccinellia distans (L.) Parl. EUROPEAN ALKALI GRASS.

Saccharum alopecuroideum (L.) Nutt. Silver Plume Grass. (*Erianthus alopecuroides* (L.) Elliott)

Schizachne purpurascens (Torr.) Swallen False Melic.

Schizachyrium littorale (Nash) E. P. Bicknell Coastal Little Bluestem. (*S. scoparium* var. *littorale* (Nash) Gould; *Andropogon littoralis* Nash, *A. scoparius* Michx. var. *littoralis* (Nash) Hitchc.)

Schizachyrium scoparium (Michx.) Nash Little Bluestem. (*Andropogon scoparius* Michx.)

*Sclerochloa dura (L.) P. Beauv. Hard Grass.

†Secale cereale L. Rye.

*Setaria faberi R. A. W. Herrm. Nodding Foxtail Grass, Giant Foxtail Grass.

*Setaria glauca (L.) P. Beauv. Yellow Foxtail Grass. (Incl. *S. pumila* (Poir.) Roem. & Schult.)

*Setaria italica (L.) P. Beauv. Foxtail Millet.

*Setaria parviflora (Poir.) Kerguélen Jointed Foxtail Grass, Perennial Foxtail Grass. (Treated by some authors as *S. geniculata* (Lam.) P. Beauv.)

*Setaria verticillata (L.) P. Beauv. Bur Foxtail Grass, Bristly Foxtail Grass.

*Setaria viridis (L.) P. Beauv. Green Foxtail Grass.

Sorghastrum nutans (L.) Nash Indian Grass.

†Sorghum bicolor (L.) Moench Sorghum. (Incl. *S. vulgare* Pers.)

*Sorghum halepense (L.) Pers. Johnson Grass.

Spartina pectinata Link Prairie Cord Grass.

Sphenopholis intermedia (Rydb.) Rydb. Slender Wedge Grass. (*S. obtusata* var. *major* (Torr.) Erdman)

Sphenopholis nitida (Biehler) Scribn. Shining Wedge Grass.

Sphenopholis obtusata (Michx.) Scribn. Prairie Wedge Grass.

Sphenopholis pensylvanica (L.) Hitchc. Swamp-oats. (*Trisetum pensylvanicum* (L.) P. Beauv. ex Roem. & Schult.)

Sporobolus compositus (Poir.) Merr. Tall Dropseed, Rough Dropseed. (Treated by some authors as *S. asper* (P. Beauv.) Kunth)

Sporobolus cryptandrus (Torr.) A. Gray Sand Dropseed.

Sporobolus heterolepis (A. Gray) A. Gray Prairie Dropseed.

Sporobolus neglectus Nash Small Rush Grass.

Sporobolus ozarkanus Fernald Ozark Rough Grass, Ozark Rush Grass. (*S. vaginiflorus* var. *ozarkanus* (Fernald) Shinners)

Sporobolus vaginiflorus (Torr. ex A. Gray) A. W. Wood Poverty Grass, Sand Grass, Sheathed Rough Grass, Sheathed Rush Grass.

Torreyochloa pallida (Torr.) G. L. Church Pale Manna Grass. (*Puccinellia pallida* (Torr.) R. T. Clausen, *Glyceria pallida* (Torr.) Trin.)

Tridens flavus (L.) Hitchc. Purpletop, Tall Redtop, Grease Grass. (*Triodia flava* (L.) Smyth)

Triplasis purpurea (Walter) Chapm. Purple Sand Grass.

*Tripsacum dactyloides (L.) L. Eastern Gama Grass.

†Triticum aestivum L. Wheat.

*Vulpia bromoides (L.) Gray Brome Fescue. (Incl. *V. dertonensis* (All.) Gola, *Festuca dertonensis* (All.) Asch. & Graebn.)

*Vulpia myuros (L.) C. C. Gmel. Rat-tail Fescue. (*Festuca myuros* L.)

Vulpia octoflora (Walter) Rydb. **var. glauca** (Nutt.) Fernald Six-weeks Fescue. (*Festuca octoflora* Walter var. *glauca* (Nutt.) Fernald, incl. var. *tenella* (Willd.) Fernald)

†Zea mays L. Corn, Indian Corn, Maize.

Zizania aquatica L. Wild Rice, Annual Wild Rice.

Order **Typhales**

■ SPARGANIACEAE. Bur-reed Family

Sparganium americanum Nutt. American Bur-reed.

Sparganium androcladum (Engelm.) Morong Keeled Bur-reed.

Sparganium chlorocarpum Rydb. Small Bur-reed, Dwarf Bur-reed. (Treated by some authors as a part of *S. emersum* Rehmann)

Sparganium eurycarpum Engelm. ex A. Gray Common Bur-reed, Giant Bur-reed.

■ TYPHACEAE. Cat-tail Family

Typha angustifolia L. Narrow-leaved Cat-tail.

Typha latifolia L. Common Cat-tail, Broad-leaved Cat-tail.

Typha × glauca Godr. (**Typha angustifolia** × **T. latifolia**) Hybrid Cat-tail, Intermediate Cat-tail.

Subclass **LILIIDAE**

Order **Liliales**

■ PONTEDERIACEAE. Water-hyacinth Family

Heteranthera dubia (Jacq.) MacMill. Water Star-grass. (*Zosterella dubia* (Jacq.) Small)

Heteranthera reniformis Ruiz & Pav. Mud-plantain.

Pontederia cordata L. Pickerelweed.

■ LILIACEAE. Lily Family

Aletris farinosa L. Unicorn-root, Colic-root.

†**Allium ampeloprasum** L. Leek. (Incl. *A. porrum* L.)

Allium canadense L. Wild Garlic, Wild Onion.

Allium cernuum Roth Nodding Wild Onion.

†**Allium sativum** L. Garlic.

†**Allium schoenoprasum** L. Chives. (Incl. var. *sibiricum* (L.) Hartm.)

Allium tricoccum Aiton
 var. burdickii Hanes White Ramps.
  (*A. burdickii* (Hanes) A. G. Jones)
 var. tricoccum Ramps, Red Ramps, Wild Leek.

Allium vineale L. Field Garlic.

Asparagus officinalis L. Garden Asparagus.

Camassia scilloides (Raf.) Cory Wild Hyacinth.

Chamaelirium luteum (L.) A. Gray Fairy-wand, Devil's-bit.

Clintonia borealis (Aiton) Raf. Bluebead-lily, Yellow Clintonia.

Clintonia umbellulata (Michx.) Morong Speckled Wood-lily, White Clintonia.

Convallaria majalis L. Lily-of-the-valley.

Erythronium albidum Nutt. White Fawn-lily, White Dogtooth-violet, White Trout-lily.

Erythronium americanum Ker Gawl. Yellow Fawn-lily, Yellow Dogtooth-violet, Yellow Trout-lily.

Erythronium rostratum W. Wolf Golden-star, Beaked Yellow Fawn-lily.

†**Galanthus nivalis** L. Snowdrop.

Hemerocallis fulva (L.) L. Orange Day-lily.

†**Hemerocallis lilioasphodelus** L. Yellow Day-lily. (Incl. *H. flava* (L.) L.)

†**Hosta lancifolia** (Thunb.) Engl. Narrow-leaved Plantain-lily, Narrow-leaved Hosta.

†**Hosta ventricosa** (Salisb.) Stearn Blue Plantain-lily.

Hypoxis hirsuta (L.) Coville Yellow Star-grass.

†**Leucojum aestivum** L. Snowflake.

Lilium canadense L. Canada Lily.

†**Lilium lancifolium** Thunb. Tiger Lily. (Incl. *L. tigrinum* Ker Gawl.)

Lilium michiganense Farw. Michigan Lily.

Lilium philadelphicum L. Wood Lily, Orange-cup Lily. (Incl. var. *andinum* (Nutt.) Ker Gawl.)

Lilium superbum L. Turk's-cap Lily.

†**Lycoris squamigera** Maxim. Magic-lily, Resurrection-lily, Surprise-lily.

Maianthemum canadense Desf. Canada Mayflower. (Incl. var. *interius* Fernald)

Medeola virginiana L. Indian Cucumber-root.

Melanthium virginicum L. Bunchflower, Virginia Bunchflower.

Melanthium woodii (J. W. Robbins ex A. W. Wood) Bodkin Ozark Bunchflower, Wood's Bunchflower, Wood's-hellebore. (*Veratrum woodii* J. W. Robbins ex A. W. Wood)

*Muscari botryoides (L.) Mill. GRAPE-HYACINTH.

†Muscari comosum (L.) Mill. TASSEL-HYACINTH.

†Narcissus poeticus L. POET'S NARCISSUS.

†Narcissus pseudonarcissus L. DAFFODIL.

Nothoscordum bivalve (L.) Britton FALSE GARLIC.

*Ornithogalum nutans L. NODDING SUMMER-SNOWFLAKE.

*Ornithogalum umbellatum L. STAR-OF-BETHLEHEM.

Polygonatum biflorum (Walter) Elliott SMOOTH SOLOMON'S-SEAL. (Incl. *P. commutatum* (Schult. & Schult. f.) A. Dietr.)

Polygonatum pubescens (Willd.) Pursh HAIRY SOLOMON'S-SEAL.

Prosartes lanuginosa (Michx.) D. Don YELLOW MANDARIN. (*Disporum lanuginosum* (Michx.) G. Nicholson)

Prosartes maculata (Buckley) A. Gray NODDING MANDARIN. (*Disporum maculatum* (Buckley) Britton)

†Scilla non-scripta (L.) Hoffmanns. & Link ENGLISH WOOD-HYACINTH, ENGLISH BLUEBELL. (*Endymion non-scriptus* (L.) Garcke; *Hyacinthoides non-scripta* (L.) Chouard ex Rothm.)

*Scilla siberica Andrews SIBERIAN SQUILL.

Smilacina racemosa (L.) Desf. SOLOMON'S-PLUME, FALSE SOLOMON'S-SEAL. (Incl. var. *cylindrata* Fernald; *Maianthemum racemosum* (L.) Link)

Smilacina stellata (L.) Desf. STARRY FALSE SOLOMON'S-SEAL. (*Maianthemum stellatum* (L.) Link)

Smilacina trifolia (L.) Desf. THREE-LEAVED FALSE SOLOMON'S-SEAL. (*Maianthemum trifolium* (L.) Sloboda)

Stenanthium gramineum (Ker Gawl.) Morong FEATHER-BELLS. (Incl. var. *robustum* (S. Watson) Fernald)

Streptopus roseus Michx. ROSE TWISTED-STALK. (Incl. var. *perspectus* Fassett)

Tofieldia glutinosa (Michx.) Pers. FALSE ASPHODEL.

Trillium cernuum L. NODDING TRILLIUM.

Trillium erectum L. PURPLE TRILLIUM, ILL-SCENTED TRILLIUM, STINKING BENJAMIN.

Trillium flexipes Raf. BENT TRILLIUM, DROOPING TRILLIUM.

Trillium grandiflorum (Michx.) Salisb. LARGE WHITE TRILLIUM, LARGE-FLOWERED TRILLIUM.

Trillium nivale Riddell SNOW TRILLIUM.

Trillium recurvatum L. C. Beck PRAIRIE WAKE-ROBIN.

Trillium sessile L. WAKE-ROBIN, TOAD-SHADE, SESSILE TRILLIUM.

Trillium undulatum Willd. PAINTED TRILLIUM.

Trillium erectum × T. flexipes

†Tulipa gesnerana L. TULIP.

Uvularia grandiflora Sm. LARGE-FLOWERED BELLWORT.

Uvularia perfoliata L. STRAWBELL, BELLWORT, PERFOLIATE BELLWORT.

Uvularia sessilifolia L. MERRY-BELLS, SESSILE-LEAVED BELLWORT.

Veratrum viride Aiton FALSE HELLEBORE, WHITE-HELLEBORE, SWAMP-HELLEBORE.

Zigadenus elegans Pursh var. glaucus (Nutt.) Cronquist WHITE WAND-LILY, DEATH CAMAS —also spelled CAMASS, WHITE CAMAS. (subsp. *glaucus* (Nutt.) Hultén, *Z. glaucus* Nutt.)

■ AGAVACEAE. AGAVE FAMILY

Agave virginica L. AMERICAN ALOE, FALSE ALOE. (*Manfreda virginica* (L.) Salisb. ex Rose)

*Yucca filamentosa L. ADAM'S-NEEDLE, SPOON-LEAVED YUCCA, YUCCA.

■ SMILACACEAE. CATBRIER FAMILY

Smilax ecirrata (Engelm. ex Kunth) S. Watson UPRIGHT CARRION-FLOWER.

Smilax glauca Walter SAWBRIER. (Incl. var. *leurophylla* S. F. Blake)

Smilax herbacea L.
var. herbacea CARRION-FLOWER.
var. lasioneura (Hook.) A. DC. PALE CARRION-FLOWER. (*S. lasioneura* Hook.)
var. pulverulenta (Michx.) A. Gray DOWNY CARRION-FLOWER. (*S. pulverulenta* Michx.)

Smilax hispida Muhl. ex Torr.   BRISTLY
GREENBRIER.   (*S. tamnoides* L. var. *hispida*
(Muhl. ex Torr.) Fernald)
Smilax illinoensis Mangaly   ILLINOIS
GREENBRIER.
Smilax rotundifolia L.   COMMON GREENBRIER.

■ DIOSCOREACEAE.   YAM FAMILY

*****Dioscorea batatas** Decne.   CHINESE YAM,
CINNAMON-VINE.
Dioscorea quaternata J. F. Gmel.   WILD YAM,
WHORLED-LEAVED YAM.
Dioscorea villosa L.   WILD YAM.

■ IRIDACEAE.   IRIS FAMILY

†**Belamcanda chinensis** (L.) DC.   BLACKBERRY-
LILY.
Iris brevicaulis Raf.   LEAFY BLUE FLAG, ZIGZAG
IRIS.
Iris cristata Aiton   DWARF CRESTED IRIS,
CRESTED DWARF IRIS.
†**Iris fulva** Ker Gawl.   COPPER IRIS.
†**Iris germanica** L.   GERMAN IRIS.
*****Iris pseudacorus** L.   YELLOW FLAG, YELLOW
WATER FLAG, FLEUR-DE-LIS.
Iris verna L.   DWARF IRIS.   (Incl. var. *smalliana*
Fernald ex M. E. Edwards)
Iris versicolor L.   NORTHERN BLUE FLAG.
Iris virginica L. **var. shrevei** (Small) E. S.
Anderson   SOUTHERN BLUE FLAG.
(*I. shrevei* Small)
Sisyrinchium albidum Raf.   PALE BLUE-EYED-
GRASS, PRAIRIE BLUE-EYED-GRASS.
Sisyrinchium angustifolium Mill.   STOUT
BLUE-EYED-GRASS, COMMON BLUE-EYED-
GRASS.
Sisyrinchium atlanticum E. P. Bicknell
ATLANTIC BLUE-EYED-GRASS.
Sisyrinchium montanum Greene   NORTHERN
BLUE-EYED-GRASS.   (Incl. var. *crebrum*
Fernald)
Sisyrinchium mucronatum Michx.   NARROW-
LEAVED BLUE-EYED-GRASS.

Order **Orchidales**

■ ORCHIDACEAE.   ORCHID FAMILY

by JOHN V. FREUDENSTEIN

Aplectrum hyemale (Muhl. ex Willd.) Torr.
PUTTY-ROOT.
Arethusa bulbosa L.   DRAGON'S-MOUTH,
ARETHUSA.
Calopogon tuberosus (L.) Britton, Sterns, &
Poggenb.   GRASS-PINK, CALOPOGON.   (Incl.
C. pulchellus R. Br. ex W. T. Aiton)
Coeloglossum viride (L.) Hartm. **var. virescens**
(Muhl. ex Willd.) Luer   LONG-BRACTED
ORCHID.   (*Habenaria viridis* (L.) R. Br. ex
W. T. Aiton, incl. var. *bracteata* (Muhl. ex
Willd.) Rchb. ex A. Gray)
Corallorhiza maculata (Raf.) Raf.   SPOTTED
CORAL-ROOT.
Corallorhiza odontorhiza (Willd.) Poir.   FALL
CORAL-ROOT, SMALL-FLOWERED CORAL-ROOT.
Corallorhiza trifida Châtel.   EARLY CORAL-
ROOT.   (Incl. var. *verna* (Nutt.) Fernald)
Corallorhiza wisteriana Conrad   SPRING
CORAL-ROOT.
Cypripedium acaule Aiton   PINK LADY'S-
SLIPPER, PINK MOCCASIN-FLOWER.
Cypripedium candidum Muhl. ex Willd.
WHITE LADY'S-SLIPPER.
Cypripedium parviflorum Salisb.   YELLOW
LADY'S-SLIPPER.
    var. parviflorum   SMALL YELLOW LADY'S-
    SLIPPER.   (*C. calceolus* L. var. *parviflorum*
    (Salisb.) Fernald)
    var. pubescens (Willd.) O. W. Knight   LARGE
    YELLOW LADY'S-SLIPPER.   (*C. calceolus* L.
    var. *pubescens* (Willd.) Correll; *C. pubescens*
    Willd.)
Cypripedium reginae Walter   SHOWY LADY'S-
SLIPPER, QUEEN'S LADY'S-SLIPPER.
Cypripedium × andrewsii A. M. Fuller (**Cypri-
pedium candidum × C. parviflorum var.
parviflorum**)   ANDREWS' LADY'S-SLIPPER.
*****Epipactis helleborine** (L.) Crantz
HELLEBORINE.
Galearis spectabilis (L.) Raf.   SHOWY ORCHIS.
(*Orchis spectabilis* L.)

Goodyera pubescens (Willd.) R. Br. ex W. T. Aiton   Downy Rattlesnake-plantain.

Goodyera tesselata Lodd.   Checkered Rattlesnake-plantain.

Hexalectris spicata (Walter) Barnhart   Crested Coral-root.

Isotria medeoloides (Pursh) Raf.   Small Whorled Pogonia.

Isotria verticillata Raf.   Whorled Pogonia.

Liparis liliifolia (L.) Rich. ex Lind.   Large Twayblade, Lily-leaved Twayblade.

Liparis loeselii (L.) Rich.   Fen Orchid, Loesel's Twayblade.

Listera cordata (L.) R. Br.   Heart-leaved Twayblade.

Malaxis unifolia Michx.   Green Adder's-mouth.

Platanthera blephariglottis (Willd.) Lindl. White Fringed Orchid, Bog White Fringed Orchid.   (*Habenaria blephariglottis* (Willd.) Hook.)

Platanthera ciliaris (L.) Lindl.   Yellow Fringed Orchid.   (*Habenaria ciliaris* (L.) R. Br. ex W. T. Aiton)

Platanthera clavellata (Michx.) Luer   Club-spur Orchid, Green Woodland Orchid. (*Habenaria clavellata* (Michx.) Spreng.)

Platanthera flava (L.) Lindl. **var. herbiola** (R. Br. ex W. T. Aiton) Luer   Tubercled Rein Orchid.   (*Habenaria flava* (L.) R. Br. var. *herbiola* (R. Br. ex W. T. Aiton) Ames & Correll)

Platanthera grandiflora (Bigelow) Lindl.   Large Purple Fringed Orchid.   (*Habenaria psycodes* (L.) Spreng. var. *grandiflora* (Bigelow) A. Gray)

Platanthera hookeri (Torr. ex A. Gray) Lindl. Hooker's Orchid.   (*Habenaria hookeri* Torr. ex A. Gray)

Platanthera hyperborea (L.) Lindl.   Tall Northern Green Orchid, Tall Northern Bog Orchid.   (*Habenaria hyperborea* (L.) R. Br. ex W. T. Aiton)

Platanthera lacera (Michx.) G. Don   Ragged Fringed Orchid.   (*Habenaria lacera* (Michx.) R. Br.)

Platanthera leucophaea (Nutt.) Lindl.   Prairie Fringed Orchid.   (*Habenaria leucophaea* (Nutt.) A. Gray)

Platanthera orbiculata (Pursh) Lindl.   Round-leaved Orchid, Large Round-leaved Orchid.   (*Habenaria orbiculata* (Pursh) Torr.)

Platanthera peramoena (A. Gray) A. Gray Purple Fringeless Orchid.   (*Habenaria peramoena* A. Gray)

Platanthera psycodes (L.) Lindl.   Small Purple Fringed Orchid.   (*Habenaria psycodes* (L.) Spreng. var. *psycodes*)

Platanthera × andrewsii (M. White) Luer (**Platanthera lacera × P. psycodes**)   Andrews' Fringed Orchid.   (*Habenaria × andrewsii* M. White)

Pogonia ophioglossoides (L.) Ker Gawl.   Rose Pogonia, Snake-mouth.

Spiranthes cernua (L.) Rich.   Nodding Ladies'-tresses.

Spiranthes lacera (Raf.) Raf. **var. gracilis** (Bigelow) Luer   Slender Ladies'-tresses. (*S. gracilis* (Bigelow) L. C. Beck)

Spiranthes lucida (H. H. Eaton) Ames   Shining Ladies'-tresses.

Spiranthes magnicamporum Sheviak   Great Plains Ladies'-tresses, Prairie Ladies'-tresses.

Spiranthes ochroleuca (Rydb.) Rydb.   Yellow Ladies'-tresses, Nodding Ladies'-tresses. (*S. cernua* var. *ochroleuca* (Rydb.) Ames)

Spiranthes ovalis Lindl. **var. erostellata** Catling Lesser Ladies'-tresses, Oval-lipped Ladies'-tresses.

Spiranthes romanzoffiana Cham.   Hooded Ladies'-tresses.

Spiranthes tuberosa Raf.   Little Ladies'-tresses.

Spiranthes vernalis Engelm. & A. Gray Narrow-leaved Ladies'-tresses, Early Ladies'-tresses.

Tipularia discolor (Pursh) Nutt.   Crane-fly Orchid.

Triphora trianthophora (Sw.) Rydb.   Three-birds Orchid.   (Incl. var. *schaffneri* Camp)

Appendix 1

STATISTICAL SUMMARY

	Native	Alien (*)	Subtotal	Alien (†)	Total
Pteridophytes					
Species	85	1	86	1	87
Interspecific hybrids	21	0	21	0	21
Gymnosperms					
Species	10	1	11	6	17
Interspecific hybrids	0	0	0	0	0
Dicotyledons					
Species	1,113	419	1,532	356	1,888
Interspecific hybrids	83	11	94	12	106
Monocotyledons					
Species	577	86	663	61	724
Interspecific hybrids	11	1	12	0	12
ALL VASCULAR PLANTS					
Species	1,785	507	2,292	424	2,716
Additional major infraspecific taxa	119	20	139	4	143
Interspecific hybrids	115	12	127	12	139
GRAND TOTALS	**2,019**	**539**	**2,558**	**440**	**2,998**

As described on p. 15 above, alien taxa marked with an asterisk (*) are naturalized in all or some part of Ohio; alien taxa marked with a dagger sign (†) have been found occasionally, but are not naturalized in the state. "Additional major infraspecific taxa" refers to those instances in which a species has more than one variety or subspecies in the Ohio flora.

95

DELETIONS

In the years since 1932, when the sixth general catalog appeared, a number of species and other taxa attributed to Ohio have been deleted from the known flora in publications by various authors. In some cases, the authors were unable to find an accurately identified Ohio specimen to justify inclusion of the taxon in the state's flora. In other cases, an accurately identified specimen was found, but it was of uncertain origin, the label data being so fragmentary or ambiguous as to make it questionable if the plant was growing outside of cultivation or if it was in fact growing in Ohio.

The taxa assembled below are deleted from the known Ohio flora, at least at the present. Most of the deletions are based on the post-1932 studies mentioned above. For these taxa, the authors who either (*a*) deleted the taxon outright, or (*b*) reported being unable to locate a specimen to support its inclusion in the flora are cited after the names of the taxa in question. Other deletions are new, originating with the authors of the several sections of this seventh general catalog. For these taxa, a brief statement of the reason for the deletion is given.

Similar lists have appeared in the past, notably that by Kellerman (1899). A few of the taxa deleted here may have to be reinstated in the future. Their geographic ranges are such that they could occur in Ohio; either existing specimens of these taxa may be discovered in herbarium collections or new populations of the plants may be found in the wild.

Pteridophytes

■ EQUISETACEAE

Equisetum × *litorale* Kühlew. ex Rupr. (*Equisetum arvense* L. × *E. fluviatile* L.) New deletion; no specimen.

Equisetum palustre L. MARSH HORSETAIL. Adams (in Cooperrider, 1982).

■ LYCOPODIACEAE

Huperzia selago (L.) Bernh. ex Schrank & Mart. NORTHERN CLUB-MOSS, NORTHERN FIR-MOSS. (*Lycopodium selago* L.) New deletion; no specimen.

Lycopodium annotinum L. BRISTLY CLUB-MOSS. New deletion; no specimen.

■ PTERIDACEAE

Cheilanthes lanosa (Michx.) D. C. Eaton HAIRY LIP FERN. New deletion; no specimen.

Gymnosperms

■ CUPRESSACEAE

Juniperus horizontalis Moench CREEPING JUNIPER. New deletion; sole specimen misidentified.

Dicotyledons

- ACERACEAE

Acer negundo L. var. *texanum* Pax   Cooperrider (1995).

Acer rubrum L. var. *drummondii* (Hook. & Arn. ex Nutt.) Sarg.   Cooperrider (1995).

Acer saccharum Marshall var. *schneckii* Rehder   New deletion (based on Braun, 1961); no specimen.

- APIACEAE or UMBELLIFERAE

Chaerophyllum tainturieri Hook.   Cooperrider (1995).

Imperatoria ostruthium L.   MASTERWORT. (*Peucedanum ostruthium* (L.) W. D. J. Koch) Cooperrider (1995).

Pimpinella saxifraga L.   BURNET-SAXIFRAGE. Cooperrider (1995).

Sanicula smallii E. P. Bicknell   Cooperrider (1995).

Thaspium pinnatifidum (Buckley) A. Gray   CUT-LEAVED MEADOW-PARSNIP.   Cooperrider (1985).

- ASCLEPIADACEAE

Asclepias floridana Lam. (*Acerates floridana* (Lam.) Hitchc.)   Cooperrider (1995).

Asclepias incarnata L. subsp. *pulchra* (Ehrh. ex Willd.) Woodson   Andreas and Cooperrider (1980).

- ASTERACEAE or COMPOSITAE

Cirsium hillii (Canby) Fernald   HILL'S THISTLE. Cusick (1997).

Eupatorium vaseyi Porter   (*E. album* L. var. *vaseyi* (Porter) Cronquist; *E. sessilifolium* L. var. *vaseyi* (Porter) Fernald & Griscom)   Roberts and Cooperrider (in Cooperrider, 1982).

Hieracium traillii Greene   SHALE BARREN HAWKWEED.   New deletion; no specimen.

Hieracium gronovii L. × *H. venosum* L.   New deletion; no specimen.

Liatris scabra (Greene) K. Schum.   SOUTHERN BLAZING-STAR.   (Treated by some authors

as a part of *L. squarrulosa* Michx.)   Fisher (1988).

Rudbeckia subtomentosa Pursh   SWEET CONEFLOWER.   New deletion; no specimen.

Senecio antennariifolius Britton   SHALE BARREN GROUNDSEL.   Roberts and Cooperrider (in Cooperrider, 1982).

Silphium integrifolium Michx.   PRAIRIE ROSINWEED.   Roberts and Cooperrider (in Cooperrider, 1982).

Silphium pinnatifidum Elliott   (*S. terebinthinaceum* Jacq. var. *pinnatifidum* (Elliott) A. Gray) Roberts and Cooperrider (in Cooperrider, 1982); Fisher (1988).

Solidago patula Muhl. ex Willd. var. *strictula* Torr. & A. Gray   Roberts and Cooperrider (in Cooperrider, 1982).

- BERBERIDACEAE

Berberis canadensis Mill.   AMERICAN BARBERRY. Loconte and Blackwell (1984); Furlow (1997).

- BETULACEAE

Carpinus caroliniana Walter subsp. *caroliniana* Furlow (1997).

- BORAGINACEAE

Amsinckia intermedia Fisch. & C. A. Mey. COASTAL FIDDLE-NECK.   Cooperrider (1995).

Heliotropium amplexicaule Vahl   CLASPING HELIOTROPE.   Cooperrider (1995).

Heliotropium tenellum (Nutt.) Torr.   SLENDER HELIOTROPE.   Cooperrider (1995).

- BRASSICACEAE or CRUCIFERAE

Lunaria rediviva L.   PERENNIAL HONESTY. Furlow (1997).

- CAESALPINIACEAE

Senna obtusifolia (L.) Irwin & Barneby   SICKLE-POD.   Furlow (1997).

■ CAPRIFOLIACEAE

Diervilla hybrida Dippel   GARDEN BUSH-HONEYSUCKLE.   Cooperrider (1995).

Lonicera caprifolium L.   ITALIAN WOODBINE.   Cooperrider (1995).

Lonicera hirsuta Eaton   HAIRY HONEYSUCKLE.   Braun (1961); Hauser (1965).

■ CARYOPHYLLACEAE

Arenaria stricta Michx. var. *texana* B. L. Rob. (*Minuartia michauxii* (Fenzl) Farw. var. *texana* (B. L. Rob.) Mattf.)   Roberts and Cooperrider (in Cooperrider, 1982).

■ CHENOPODIACEAE

Atriplex argentea Nutt.   SILVERSCALE, SALTBUSH.   Arbak and Blackwell (1982); Furlow (1997).

■ CISTACEAE

Lechea stricta Legg. ex Britton   PRAIRIE PINWEED.   Roberts and Cooperrider (in Cooperrider, 1982).

■ CLETHRACEAE

Clethra alnifolia L.   SWEET PEPPERBUSH, SUMMERSWEET.   Roberts and Cooperrider (in Cooperrider, 1982); Cooperrider (1995).

■ CORNACEAE

Cornus foemina Mill. subsp. *foemina* or var. *foemina*   Cooperrider (1980a, 1995).

■ CUCURBITACEAE

Cucurbita lagenaria L.   GOURD.   (*Lagenaria siceraria* (Molina) Standl.)   Cooperrider (1995).

■ CUSCUTACEAE

Cuscuta indecora Choisy   PRETTY DODDER.   Cooperrider (1995).

■ DIPSACACEAE

Scabiosa atropurpurea L.   Hauser (1963).

■ ERICACEAE

Gaylussacia frondosa (L.) Torr. & A. Gray ex Torr.   DANGLEBERRY.   Braun (1961).

Rhododendron canescens (Michx.) Sweet   HOARY AZALEA.   Braun (1961).

Rhododendron viscosum (L.) Torr.   CLAMMY AZALEA, SWAMP AZALEA.   Braun (1961).

■ EUPHORBIACEAE

Euphorbia exigua L.   DWARF SPURGE.   Cooperrider (1995).

Euphorbia glyptosperma Engelm.   (*Chamaesyce glyptosperma* (Engelm.) Small)   Roberts and Cooperrider (in Cooperrider, 1982); Cooperrider (1995).

■ FABACEAE or PAPILIONACEAE

Apios priceana B. L. Rob.   Roberts and Cooperrider (in Cooperrider, 1982).

Desmodium lineatum DC.   MATTED TICK-TREFOIL.   Furlow (1997).

Lespedeza stuevei Nutt.   Roberts and Cooperrider (in Cooperrider, 1982).

Lespedeza hirta (L.) Hornem. × *L. procumbens* Michx.   Furlow (1997).

■ FAGACEAE

Quercus ilicifolia Wangenh.   BEAR OAK.   Braun (1961).

Quercus michauxii Nutt.   SWAMP CHESTNUT OAK.   Furlow (1997).

Quercus phellos L.   WILLOW OAK.   Braun (1961).

Quercus prinoides Willd.   DWARF CHINQUAPIN OAK.   Braun (1961); Furlow (1997).

Quercus × *fernowii* Trel. (*Quercus alba* L. × *Q. stellata* Wangenh.) Furlow (1997).

■ FUMARIACEAE

Fumaria parviflora Lam.   Furlow (1997).

■ GROSSULARIACEAE

Ribes lacustre (Pers.) Poir.   SWAMP CURRANT,
    BRISTLY BLACK CURRANT.   Furlow (1997).
Ribes rotundifolium Michx.   APPALACHIAN
    GOOSEBERRY.   New deletion; no specimen.

■ HALORAGACEAE

Myriophyllum alternifolium DC.   Stuckey and
    Roberts (1997).

■ LAMIACEAE or LABIATAE

Coleus blumei Benth.   PAINTED-NETTLE.
    Cooperrider (1995).
Lamium album L.   SNOWFLAKE.   Cooperrider
    (1995).
Mentha arvensis L. var. *arvensis*   Cooperrider
    (1995).
Pycnanthemum torrei Benth.   Cooperrider (1995).
Salvia urticifolia L.   Roberts and Cooperrider (in
    Cooperrider, 1982); Cooperrider (1995).
Teucrium botrys L.   CUT-LEAVED GERMANDER.
    Cooperrider (1995).
Teucrium scorodonia L.   WOOD GERMANDER.
    Cooperrider (1995).
Thymus serpyllum L.   WILD THYME.
    Cooperrider (1995).

■ LENTIBULARIACEAE

Utricularia ochroleuca R. W. Hartm.   New dele-
    tion; no specimen.

■ LOGANIACEAE

Spigelia marilandica (L.) L.   INDIAN-PINK,
    PINKROOT.   Roberts and Cooperrider (in
    Cooperrider, 1982); Cooperrider (1995).

■ LYTHRACEAE

Didiplis diandra (Nutt. ex DC.) A. W. Wood
    WATER-PURSLANE.   (*Peplis diandra* Nutt. ex

DC.)   Stuckey and Roberts (1977); Cooper-
    rider (1995).

■ MALVACEAE

Callirhoe involucrata (Torr. & A. Gray) A. Gray
    POPPY-MALLOW.   Cooperrider (1995).

■ MORACEAE

Morus nigra L.   BLACK MULBERRY.   New dele-
    tion; no specimen.

■ MYRICACEAE

Myrica gale L.   SWEET GALE.   New deletion; no
    specimen.

■ ONAGRACEAE

Oenothera albicaulis Pursh   PRAIRIE EVENING-
    PRIMROSE.   Cooperrider (1995).

■ PLUMBAGINACEAE

Ceratostigma plumbaginoides Bunge   Cooperrider
    (1995).

■ POLEMONIACEAE

Phlox glaberrima L. var. *glaberrima*, var. *interior*
    Wherry, and var. *melampyrifolia* (Salib.)
    Wherry   Cooperrider (1986).

■ PORTULACACEAE

Claytonia perfoliata Donn ex Willd.   Furlow (1997).

■ RANUNCULACEAE

Thalictrum clavatum DC.   LADY-RUE.   Furlow
    (1997).

■ RHAMNACEAE

Berchemia scandens (Hill) K. Koch   SUPPLE-JACK.
    Cooperrider (1995).

ROSACEAE

Crataegus prunifolia Pers.   Furlow (1997).

Crataegus × mansfieldensis Sarg.   (*Crataegus coccinea* L. × *C. punctata* Jacq.)   MANSFIELD HAWTHORN.   Furlow (1997).

Crataegus chrysocarpa Ashe × *C. mollis* Scheele   Furlow (1997).

Prunus angustifolia Marshall   CHICKASAW PLUM.   New deletion; no specimen.

Pyrus arbutifolia (L.) L. f.   RED CHOKEBERRY.   (*Aronia arbutifolia* (L.) Pers.)   Braun (1961).

Rosa virginiana Mill.   VIRGINIA ROSE.   Braun (1961).

Spiraea latifolia (Aiton) Borkh.   NORTHERN MEADOW-SWEET.   (*S. alba* Du Roi var. *latifolia* (Aiton) Dippel)   New deletion; sole specimen of uncertain origin.

RUBIACEAE

Galium latifolium Michx.   WIDE-LEAVED BEDSTRAW.   Hauser (1964).

Galium vernum Scop.   Cooperrider (1995).

Houstonia tenuifolia Nutt.   Hauser (1964); Cooperrider (1995).

SALICACEAE

Salix cordata Michx.   SAND-DUNE WILLOW.   (Incl. *S. syrticola* Fernald)   New deletion; no correctly identified specimen.

Salix × blanda Andersson   (*Salix babylonica* L. × *S. fragilis* L.)   BLAND'S WILLOW.   Furlow (1997).

SANTALACEAE

Geocaulon lividum (Richardson) Fernald (*Comandra livida* Richardson)   Roberts and Cooperrider (in Cooperrider, 1982); Furlow (1997).

SAXIFRAGACEAE

Mitella nuda L.   NAKED MITREWORT, NAKED BISHOP'S-CAP.   Cooperrider (1980b).

SCROPHULARIACEAE

Penstemon brevisepalus Pennell   SHORT-SEPALED BEARD-TONGUE.   Cooperrider (1976, 1995).

SOLANACEAE

Datura metel L.   HORN-OF-PLENTY.   Cooperrider (1995).

Solanum elaeagnifolium Cav.   SILVER-LEAVED NIGHTSHADE.   Cooperrider (1995).

Solanum pseudocapsicum L.   JERUSALEM-CHERRY.   Cooperrider (1995).

Solanum sisymbriifolium Lam.   STICKY NIGHTSHADE.   Cooperrider (1995).

TILIACEAE

Tilia floridana Small   Roberts and Cooperrider (in Cooperrider, 1982).

ULMACEAE

Ulmus alata Michx.   WINGED ELM.   Braun (1961); Furlow (1997).

Ulmus serotina Sarg.   SEPTEMBER ELM.   Braun (1961).

URTICACEAE

Urtica urens L.   BURNING NETTLE, DOG NETTLE.   Furlow (1997).

VERBENACEAE

Verbena × hybrida Voss ex Rümpler   GARDEN VERVAIN, GARDEN VERBENA.   Cooperrider (1995).

VIOLACEAE

Viola villosa Walter   SOUTHERN WOOLLY VIOLET.   Miller (cited in Cooperrider, 1982, 1995.)

Monocotyledons

COMMENLINACEAE

Commelina erecta L.   ERECT DAYFLOWER.   New deletion; no specimen.

■ CYPERACEAE

Carex baileyi Britton   Bailey's Sedge.   Cusick
 (1996).
Carex canescens L. var. *canescens*   New deletion;
 no specimen.
Carex cumulata (L. H. Bailey) Fernald   Braun
 (1967).
Carex gravida L. H. Bailey   Heavy Sedge.
 Cusick (1996).
Carex ormostachya Wiegand   (Treated by some
 authors as part of *C. gracilescens* Steud.)
 Cusick (1996).
Carex paupercula Michx. var. *pallens* Fernald
 Braun (1967).
Carex albursina E. Sheld. × *C. careyana* Torr. ex
 Dewey   New deletion; no specimen.
Cyperus houghtonii Torr.   Houghton's
 Umbrella-sedge.   New deletion; no
 specimen.
Scirpus olneyi A. Gray (Treated by some authors
 as a part of *S. americanus* Pers.)   Braun
 (1967).

■ IRIDACEAE

Iris lacustris Nutt.   Dwarf Lake Iris.   Braun
 (1967).

■ JUNCACEAE

Juncus scirpoides Lam.   Braun (1967).
Juncus vaseyi Engelm.   Vasey's Rush.   New
 deletion; no specimen.

■ MAYACACEAE

Mayaca aubletii Michx.   Gleason (1952); Stuckey
 and Roberts (1977).

■ ORCHIDACEAE

Goodyera repens (L.) R. Br. ex W. T. Aiton var.
 ophioides Fernald   Dwarf Rattlesnake-
 plantain.   Stuckey and Roberts (in
 Cooperrider, 1982).

■ POACEAE or GRAMINEAE

Echinochloa colonum (L.) Link   Jungle-rice.
 Weishaupt (in Braun, 1967).
Panicum malacophyllum Nash   Soft-leaved
 Panic Grass.   (*Dicanthelium malacophyllum*
 (Nash) Gould)   New deletion; no specimen.
Panicum villosissimum Nash var. *villosissimum*
 Villous Panic Grass.   (Incl. var. *pseudopu-
 bescens* (Nash) Fernald; *Dicanthelium villosissi-
 mum* (Nash) Freckmann var. *villosissimum*)
 New deletion; no specimen.
Poa autumnalis Muhl. ex Elliott   Weishaupt (in
 Braun, 1967).
Sporobolus clandestinus (Biehler) Hitchc.
 Weishaupt (in Braun, 1967).

■ RUPPIACEAE

Ruppia maritima L.   Ditch-grass.   New deletion; Ohio plants now assigned to *R. cirrhosa*.

■ SMILACACEAE

Smilax bona-nox L.   Fringed Greenbrier.
 Braun (1961, 1967).

■ XYRIDACEAE

Xyris caroliniana Walt.   Carolina Yellow-eyed-
 grass.   New deletion; Ohio plants formerly
 identified as this species now assigned to
 X. difformis.

LITERATURE CITED

Andreas, B. K., and T. S. Cooperrider. 1980. The Asclepiadaceae of Ohio. Castanea 45:51–55.

Arbak, Z. E., and W. H. Blackwell. 1982. The Chenopodiaceae of Ohio. Castanea 47:284–297.

Bailey Hortorium Staff. 1976. Hortus Third. Macmillan Publishing Co., New York. xiv + 1290 pp.

Barkworth, M. E., K. M. Capels, and L. A. Vorobik, eds. In prep. Manual of grasses for the continental United States and Canada.

Beardslee, H. C. 1874. Catalogue of the plants of Ohio, including flowering plants, ferns, mosses and liverworts. Painesville, OH. 19 pp.

Braun, E. L. 1950. Deciduous forests of eastern North America. Blakiston Co., Philadelphia. xiv + 596 pp.

———. 1961. The woody plants of Ohio. Ohio State University Press, Columbus. 362 pp.

———. 1967. The Monocotyledoneae [of Ohio]. Cat-tails to orchids. With the Gramineae by Clara G. Weishaupt. Ohio State University Press, Columbus. ix + 464 pp.

Brummitt, R. K., and C. E. Powell, eds. 1992. Authors of plant names. Royal Botanic Gardens, Kew. 732 pp.

Cooperrider, T. S. 1961. Ohio floristics at the county level. Ohio J. Sci. 61:318–320.

———. 1976. Notes on Ohio Scrophulariaceae. Castanea 41:223–226.

———. 1980a. *Cornus foemina* excluded from the known Ohio flora. Castanea 45:216–217.

———. 1980b. *Mitella nuda* excluded from the known Ohio flora. Castanea 45:282–283.

———, ed. 1982. Endangered and threatened plants of Ohio. Ohio Biol. Survey Biol. Notes,
No. 16. The Ohio State University, Columbus. 92 pp.

———. 1984. Ohio's herbaria and the Ohio Flora Project. Ohio J. Sci. 84:189–196.

———. 1985. *Thaspium* and *Zizia* (Umbelliferae) in Ohio. Castanea 50:116–119.

———. 1986. The genus *Phlox* (Polemoniaceae) in Ohio. Castanea 51:145–148.

———. 1992. Changes in knowledge of the vascular plant flora of Ohio, 1860–1991. Ohio J. Sci. 92:73–76.

———. 1995. The Dicotyledoneae of Ohio. Part 2. Linaceae through Campanulaceae. Ohio State University Press, Columbus. xxi + 656 pp.

———. 1999. On the flora of Ohio. Denison J. Biol. Sci. 33:11–14.

Cronquist, A. 1981. An integrated system of classification of flowering plants. Columbia University Press, New York. xviii + 1262 pp.

Cusick, A. W. 1996. Notes on selected species of *Carex* (Cyperaceae) in Ohio. Castanea 61:382–390.

———. 1997. *Cirsium hillii* deleted from the Ohio Flora. Abstract. Ohio J. Sci. 97(2): A-23–A-24.

Cusick, A. W., and J. A. Snider, eds. 1982. Survey of the herbarium resources of Ohio. Organization of Herbaria in Ohio, Columbus. 43 pp.

Dean, F. W., and L. C. Chadwick. 1940. Ohio trees. The Ohio State University, Columbus. 104 pp.

Fernald, M. L. 1950. Gray's manual of botany. 8th ed. American Book Co., New York. lxiv + 1632 pp.

Fisher, T. R. 1988. The Dicotyledoneae of Ohio.

Part 3. Asteraceae. Ohio State University Press, Columbus. x + 280 pp.

Flora of North America Editorial Committee, eds. Flora of North America north of Mexico. Oxford University Press, New York and Oxford.
> 1993a. Vol. 1. Introduction. xxi + 372 pp.
> 1993b. Vol. 2. Pteridophytes and Gymnosperms. xvi + 475 pp.
> 1997. Vol. 3. Magnoliophyta: Magnoliidae and Hamamelidae. xxiii + 590 pp.

Furlow, J. J. 1997. The vascular flora of Ohio. Vol. 2, Part 1. Dicotyledoneae: Saururaceae through Fabaceae. Checklist. 2nd ed. Processed and distributed by author. 107 pp.
———. In prep. The Dicotyledoneae of Ohio. Part 1. Saururaceae through Fabaceae.

Gleason, H. A. 1952. The new Britton and Brown illustrated flora of the northeastern United States and adjacent Canada. Vol. 1. New York Botanical Garden, Bronx. lxxv + 482 pp.

Gleason, H. A., and A. Cronquist. 1991. Manual of vascular plants of northeastern United States and adjacent Canada. 2nd ed. New York Botanical Garden, Bronx. lxxv + 910 pp.

Gordon, R. B. 1966. Natural vegetation of Ohio at the time of the earliest land surveys. Map. Ohio Biological Survey, Columbus.
———. 1969. The natural vegetation of Ohio in pioneer days. Ohio Biol. Surv. Bull. N.S. 3:i-xi, 1–113.

Hauser, E. J. P. 1963. The Dipsacaceae and Valerianaceae of Ohio. Ohio J. Sci. 63:26–30.
———. 1964. The Rubiaceae of Ohio. Ohio J. Sci. 64:27–35.
———. 1965. The Caprifoliaceae of Ohio. Ohio J. Sci. 65:118–129.

Kartesz, J. T. 1994. A synonymized checklist of the vascular flora of the United States, Canada, and Greenland. 2nd ed. Timber Press, Portland, Oregon.
> Vol. 1. Checklist. lxi + 622 pp.
> Vol. 2. Thesaurus. vii + 816 pp.

Kartesz, J. T., and C. A. Meacham. 1999. Synthesis of the North American flora, CD-ROM, version 1.0. North Carolina Botanical Garden, University of North Carolina, Chapel Hill.

Kellerman, W. A. 1899. The fourth state catalogue of Ohio plants, consisting of a serially numbered systematic check list of the pteridophytes and spermatophytes. Bull. Ohio State Univ., Ser. 4. 10:1–65.

Kellerman, W. A., and W. C. Werner. 1893. Catalogue of Ohio plants. Rep. Geol. Surv. Ohio 7:56–406.

Knepper, G. W. 1976. An Ohio portrait. Ohio Historical Soc., Columbus. i + 282 pp.

Lafferty, M. B., ed. 1979. Ohio's natural heritage. The Ohio Academy of Science, Columbus. 324 pp.

Loconte, H., and W. H. Blackwell, Jr. 1984. Berberidaceae of Ohio. Castanea 49:39–43.

Lowden, R. M. 1997. Riddell, Sullivant, and the early botanical exploration of Franklin County, Ohio, U.S.A. (1832–1840). Taxon 46:689–703.

McCormac, J. S. 1999. Best plant finds of 1998. Ohio Div. Natural Areas and Preserves Newsletter 21:2,8.

Moseley, R. E., Jr. 1990. Division of Natural Areas and Preserves, p. 197–208. In C. C. King, ed., A legacy of stewardship: The Ohio Department of Natural Resources, 1949–1989. Ohio Dept. Natural Resources, Columbus. xv + 280 pp.

Newberry, J. S. 1860. Catalogue of the flowering plants and ferns of Ohio. Ann. Rep. Ohio State Board Agric. 14:235–273.

Ohio Division of Natural Areas and Preserves. 1996. Directory of Ohio's state nature preserves. Ohio Dept. Natural Resources, Columbus. xii + 114 pp.
———. 1998. Rare native Ohio plants: 1998–99 status list. Ohio Dept. Natural Resources, Columbus. 29 pp.

Roberts, M. L., and R. L. Stuckey. 1974. Bibliography of theses and dissertations on Ohio floristics and vegetation in Ohio colleges and universities. Ohio Biol. Surv. Inform. Circ. No. 7. 92 pp.

Schaff, M. 1905. Etna and Kirkersville. Houghton, Miflin and Co., Boston and New York. 157 pp.

Schaffner, J. H. 1914. Catalog of Ohio vascular plants. Ohio Biol. Surv. Bull. 2:125–247.
———. 1932. Revised catalog of Ohio vascular plants. Ohio Biol. Survey Bull. 25:87–215.

Stuckey, R. L. 1982. Historical review, p. 5–9. In T. S. Cooperrider, ed., Endangered and threatened plants of Ohio. Ohio Biol. Surv. Biol. Notes, No. 16. The Ohio State University, Columbus. 92 pp.

————. 1984. Early Ohio botanical collections and the development of the State Herbarium. Ohio J. Sci. 84:148–174.

Stuckey, R. L., and M. L. Roberts. 1977. Rare and endangered aquatic vascular plants of Ohio: an annotated list of the imperiled species. Sida 7:24–41.

Vincent, M. A., and A. W. Cusick. 1998. New records of alien species in the Ohio vascular flora. Ohio J. Sci. 98:10–17.

Weishaupt, C. G. 1971. Vascular plants of Ohio: a manual for use in field and laboratory. 3rd ed. Kendall/Hunt Publishing Co., Dubuque. iii + 292 pp.

INDEX TO SCIENTIFIC NAMES

The primary names used in the catalog text are printed below in regular (roman) type. The epithets of synonyms and those of other alternate names, given in *italics* in the text, are in *italics* below. Epithets of the names of deleted taxa (Appendix 2) also are in *italics*. Family names are in CAPITALS. Names occurring in the introductory sections are not indexed.

graveolens, 54
Angelica
  atropurpurea, 54
  venenosa, 54
ANNONACEAE, 23
Anoda
  cristata, 33
Antennaria
  howellii, 67
    subsp. *canadensis*, 67
    subsp. *neodioica*, 67
    subsp. *petaloidea*, 67
  neglecta, 67
    var. canadensis, 67
    var. neglecta, 67
    var. neodioica, 67
    var. petaloidea, 67
  neodioica, 67
    subsp. *canadensis*, 67
    subsp. *neodioica*, 67
    subsp. *petaloidea*, 67
  parlinii, 67
    subsp. *falax*, 67
    subsp. *parlinii*, 67
  plantaginifolia, 67
    var. ambigens, 67
    var. parlinii, 67
    var. plantaginifolia, 67
  solitaria, 67
  virginica, 67
Anthemis
  arvensis, 67
  cotula, 67
  nobilis, 67
  tinctoria, 67
Anthoxanthum
  aristatum, 84
  odoratum, 84
Anthriscus
  caucalis, 54
  sylvestris, 54
Anthyllis
  vulneraria, 46
Antirrhinum
  majus, 62
  orontium, 62
Apera
  interrupta, 84
  spica-venti, 84
APIACEAE, 54, 97
Apios
  americana, 46
  priceana, 98
Apium
  graveolens, 54
Aplectrum

hyemale, 92
APOCYNACEAE, 55
Apocynum
  androsaemifolium, 55
  cannabinum, 55
    var. *hypericifolium*, 55
  × floribundum, 55
  × *medium*, 55
  sibiricum, 55
    var. *cordigerum*, 55
  androsaemifolium × cannabinum, 55
AQUIFOLIACEAE, 50
Aquilegia
  canadensis, 24
  vulgaris, 24
Arabidopsis
  thaliana, 37
Arabis
  canadensis, 37
  divaricarpa, 37
  drummondii, 37
  glabra, 37
  hirsuta, 37
    var. adpressipilis, 37
    var. pycnocarpa, 37
  laevigata, 37
  lyrata, 37
  patens, 37
  perstellata, 37
    var. *shortii*, 37
  shortii, 37
ARACEAE, 77
Arales, 77
Aralia
  elata, 54
  hispida, 54
  nudicaulis, 54
  racemosa, 54
  spinosa, 54
ARALIACEAE, 54
Araliales, 54
Arctium
  lappa, 67
  minus, 67
Arctostaphylos
  uva-ursi, 39
    var. *coactilis*, 39
Arecidae, 77
Arenaria
  lateriflora, 30
  patula, 30
  serpyllifolia, 30
  stricta, 30
    var. *texana*, 98
Arethusa

bulbosa, 92
Argemone
  albiflora, 26
  mexicana, 26
Argentina
  anserina, 44
Arisaema
  atrorubens, 77
  dracontium, 77
  stewardsonii, 77
  triphyllum, 77
    var. *pusillum*, 77
    var. *stewardsonii*, 77
Aristida
  dichotoma, 84
  longespica, 84
    var. geniculata, 84
  necopina, 84
  oligantha, 84
  purpurascens, 84
Aristolochia
  clematitis, 23
  serpentaria, 23
  tomentosa, 23
ARISTOLOCHIACEAE, 23
Aristolochiales, 23
Armoracia
  aquatica, 37
  lacustris, 37
  rusticana, 37
Arnoglossum
  atriplicifolium, 67
  muhlenbergii, 67
  plantagineum, 67
Arnoseris
  minima, 67
Aronia
  arbutifolia, 100
  floribunda, 44
  melanocarpa, 44
  prunifolia, 44
Arrhenatherum
  elatius, 84
Artemisia
  absinthium, 67
  annua, 67
  biennis, 67
  campestris, 67
    var. canadensis, 67
    var. caudata, 67
  canadensis, 67
  caudata, 67
  gmelinii, 68
  ludoviciana, 68
    var. *gnaphalodes*, 68
  pontica, 68

BUXACEAE, 50
Buxus
  sempervirens, 50

Cabomba
  caroliniana, 24
CABOMBACEAE, 24
Cacalia
  atriplicifolia, 67
  muhlenbergii, 67
  plantaginea, 67
  suaveolens, 71
  tuberosa, 67
CACTACEAE, 29
CAESALPINIACEAE, 45,97
Cakile
  eduntula, 37
    subsp. *lacustris*, 37
    var. lacustris, 37
Calamagrostis
  canadensis, 84
  cinnoides, 84
  coarctata, 84
  inexpansa, 84
  insperata, 84
  porteri, 84
    subsp. insperata, 84
  stricta, 84
    subsp. *inexpansa*, 84
Calamintha
  arkansana, 59
Calamovilfa
  longifolia, 84
    var. magna, 84
Calendula
  officinalis, 69
Calla
  palustris, 77
Callirhoe
  involucrata, 99
CALLITRICHACEAE, 61
Callitrichales, 61
Callitriche
  heterophylla, 61
  palustris, 61
  terrestris, 61
  verna, 61
Calluna
  vulgaris, 39
Calopogon
  pulchellus, 92
  tuberosus, 92
Caltha
  palustris, 24
CALYCANTHACEAE, 23
Calycanthus

fertilis, 23
floridus, 23
  var. glaucus, 23
Calystegia
  hederacea, 57
  pellita, 57
  pubescens, 57
  sepium, 57
  spithamaea, 57
Camassia
  scilloides, 90
Camelina
  microcarpa, 37
  sativa, 37
Campanula
  americana, 64
  aparinoides, 64
    var. aparinoides, 64
    var. grandiflora, 64
  rapunculoides, 64
  rotundifolia, 64
  trachelium, 64
  uliginosa, 64
CAMPANULACEAE, 64
Campanulales, 64
Campanulastrum
  americanum, 64
Campsis
  radicans, 64
Camptosorus
  rhizophyllus, 19
CANNABACEAE, 26
Cannabis
  sativa, 26
CAPPARACEAE, 37
Capparales, 37
CAPRIFOLIACEAE, 65, 98
Capsella
  bursa-pastoris, 37
Cardamine
  angustata, 38
  bulbosa, 38
  concatenata, 38
  diphylla, 38
  dissecta, 38
  douglassii, 38
  flexuosa, 38
  hirsuta, 38
  impatiens, 38
  parviflora, 38
    var. arenicola, 38
  pensylvanica, 38
  pratensis, 38
    var. palustris, 38
    var. pratensis, 38
  rhomboidea, 38

rotundifolia, 38
Cardaria
  draba, 38
Cardiospermum
  halicacabum, 52
Carduus
  acanthoides, 69
  nutans, 69
Carex
  abscondita, 78
  aggregata, 78
  alata, 78
  albicans, 78
    var. albicans, 78
    var. emmonsii, 78
  albolutescens, 78
  albursina, 78
  alopecoidea, 78
  amphibola, 78
    var. amphibola, 78
    var. rigida, 78
    var. turgida, 79
  annectens, 79
    var. *xanthocarpa*, 79
  appalachica, 79
  aquatilis, 79
    var. *altior*, 79
    var. substricta, 79
  arctata, 79
  argyrantha, 79
  artitecta, 78
  atherodes, 79
  atlantica, 79
    subsp. *atlantica*, 79
    subsp. *capillacea*, 79
    var. atlantica, 79
    var. capillacea, 79
  aurea, 79
  baileyi, 101
  bebbii, 79
  bicknellii, 79
  blanda, 79
  brachyglossa, 79
  brevior, 79
  bromoides, 79
  brunnescens, 79
    subsp. *sphaerostachya*, 79
    var. sphaerostachya, 79
  bushii, 79
  buxbaumii, 79
  canescens, 79
    subsp. *disjuncta*, 79
    var. *canescens*, 101
    var. disjuncta, 79
  careyana, 79
  caroliniana, 79

Corallorhiza
  maculata, 92
  odontorhiza, 92
  trifida, 92
    var. *verna*, 92
  wisteriana, 92
Coreopsis
  grandiflora, 69
  lanceolata, 70
  major, 70
  tinctoria, 70
  tripteris, 70
  verticillata, 70
Coriandrum
  sativum, 54
Corispermum
  americanum, 29
  hyssopifolium, 29
    var. *americanum*, 29
  nitidum, 29
  pallasii, 29
CORNACEAE, 50, 98
Cornales, 50
Cornus
  alternifolia, 50
  amomum, 50
    subsp. *amomum*, 50
    subsp. *obliqua*, 50
    var. amomum, 50
    var. schuetzeana, 50
  × arnoldiana, 50
  canadensis, 50
  drummondii, 50
  florida, 50
  foemina, 98
    subsp. *foemina*, 98
    var. *foemina*, 98
  obliqua, 50
  racemosa, 50
  rugosa, 50
  sericea, 50
  stolonifera, 50
    var. *baileyi*, 50
  amomum × racemosa, 50
Coronilla
  varia, 46
Coronopus
  didymus, 38
Corydalis
  aurea, 26
  flavula, 26
  sempervirens, 26
Corylus
  americana, 28
  cornuta, 28
Cosmos

bipinnatus, 70
Cotinus
  coggygria, 53
Cotoneaster
  divaricatus, 42
  pyracantha, 44
  simonsii, 42
CRASSULACEAE, 41
Crataegus
  × anomala, 43
  arborea, 42
  basilica, 42
  beata, 42
  biltmoreana, 42
  boyntonii, 42
  brainerdii, 42
    var. *brainerdii*, 42
    var. *scabrida*, 42
  brumalis, 42
  calpodendron, 42
    var. *calpodendron*, 42
    var. *globosa*, 42
    var. *microcarpa*, 42
  × chadsfordiana, 43
  chrysocarpa, 42
  coccinea, 42
  coleae, 42
  compacta, 43
  crawfordiana, 43
  crus-galli, 42
    var. *barrettiana*, 42
    var. *crus-galli*, 42
    var. *exigua*, 42
    var. *leptophylla*, 42
    var. *pachyphylla*, 42
    var. *pyracanthifolia*, 42
  disjuncta, 43
  × disperma, 43
  engelmannii, 42
  flabellata, 42
  fontanesiana, 42
  formosa, 43
  fortunata, 42
  franklinensis, 43
  gattingeri, 43
  gaudens, 43
  gravis, 42
  habereri, 42
  hannibalensis, 42
  hillii, 42
  holmesiana, 42
  horseyi, 42
  × hudsonica, 43
  indicens, 43
  intricata, 42
    var. *intricata*, 42

var. *straminea*, 42
iracunda, 42
  var. *silvicola*, 42
jesupii, 43
× kellermanii, 43
× *laetifica*, 43
laevigata, 42
leiophylla, 43
× locuples, 43
× lucorum, 43
mackenzii, 43
  var. *bracteata*, 43
macrosperma, 42
  var. *acutiloba*, 42
  var. *demissa*, 42
  var. *macrosperma*, 42
  var. *matura*, 42
  var. *pentranda*, 42
  var. *roanensis*, 42
× mansfieldensis, 100
margarettiae, 42
  var. *brownii*, 42
  var. *margarettiae*, 42
  var. *meiophylla*, 42
milleri, 43
mollis, 42
  var. *mollis*, 42
  var. *sera*, 42
  var. *submollis*, 42
monogyna, 42
ohioensis, 42
oxyacantha, 42
pedicellata, 42
  var. *albicans*, 42
  var. *assurgens*, 42
  var. *pedicellata*, 42
  var. *robesoniana*, 42
pennsylvanica, 42
peoriensis, 43
× persimilis, 43
phaenopyrum, 42
populnea, 42
porteri, 43
pringlei, 42
pruinosa, 42
  var. *dissona*, 43
  var. *latisepala*, 43
  var. *pruinosa*, 43
prunifolia, 100
punctata, 43
  var. *aurea*, 43
  var. *canescens*, 43
  var. *microphylla*, 43
  var. *pausiaca*, 43
  var. *punctata*, 43
putnamiana, 42

Hystrix
  patula, 85

Iberis
  umbellata, 38
Ilex
  opaca, 50
  verticillata, 50
    var. *padifolia*, 50
    var. *tenuifolia*, 50
Impatiens
  balsamina, 54
  capensis, 54
  pallida, 54
Imperatoria
  ostruthium, 97
Inula
  helenium, 72
Iodanthus
  pinnatifidus, 38
Ionactis
  linariifolius, 68
Ipomoea
  coccinea, 57
  hederacea, 57
  lacunosa, 57
  pandurata, 57
  purpurea, 57
Ipomopsis
  rubra, 57
IRIDACEAE, 92, 101
Iris
  brevicaulis, 92
  cristata, 92
  fulva, 92
  germanica, 92
  lacustris, 101
  pseudacorus, 92
  shrevei, 92
  verna, 92
    var. *smalliana*, 92
  versicolor, 92
  virginica, 92
    var. shrevei, 92
Isanthus
  brachiatus, 59
ISOETACEAE, 17
Isoetales, 17
Isoetes
  echinospora, 17
  engelmannii, 18
Isoetopsida, 17
Isopyrum
  biternatum, 24
Isotria
  medeoloides, 93

verticillata, 93
Iva
  annua, 72
  xanthifolia, 72

Jacquemontia
  tamnifolia, 57
Jeffersonia
  diphylla, 25
JUGLANDACEAE, 27
Juglandales, 27
Juglans
  cinerea, 27
  nigra, 27
JUNCACEAE, 77, 101
JUNCAGINACEAE, 76
Juncales, 77
Juncus
  acuminatus, 77
  alpinoarticulatus, 77
  alpinus, 77
    var. *fuscescens*, 77
    var. *rariflorus*, 77
  anthelatus, 78
  arcticus, 78
    var. *balticus*, 78
    var. *littoralis*, 78
  articulatus, 78
  balticus, 78
  biflorus, 78
  brachycarpus, 78
  brachycephalus, 78
  bufonius, 78
  canadensis, 78
  compressus, 78
  dichotomus, 78
  diffusissimus, 78
  dudleyi, 78
  effusus, 78
    var. *decipiens*, 78
    var. *pylaei*, 78
    var. *solutus*, 78
  gerardii, 78
  greenei, 78
  interior, 78
  marginatus, 78
  nodosus, 78
  platyphyllus, 78
  scirpoides, 101
  secundus, 78
  × stuckeyi, 78
  subcaudatus, 78
  tenuis, 78
    var. *anthelatus*, 78
    var. *dudleyi*, 78
    var. *platyphyllus*, 78

var. tenuis, 78
    var. *williamsii*, 78
  torreyi, 78
  vaseyi, 101
  alpinoarticulatus × torreyi, 78
Juniperus
  communis, 22
    subsp. *depressa*, 22
    var. depressa, 22
  horizontalis, 96
  virginiana, 22
    var. *crebra*, 22
Justicia
  americana, 64

Kalmia
  latifolia, 39
Kerria
  japonica, 43
Kickxia
  elatine, 62
  spuria, 62
Kochia
  scoparia, 29
    var. *culta*, 29
Koeleria
  cristata, 86
  macrantha, 86
  pyramidata, 86
Koelreuteria
  paniculata, 52
Krigia
  biflora, 72
  dandelion, 72
  virginica, 72
Kuhnia
  eupatorioides, 72
    var. *corymbulosa*, 72
Kummerowia
  stipulacea, 46
  striata, 46
Kyllinga
  pumila, 82
  tenuifolia, 82

LABIATAE, 59, 99
Lablab
  purpureus, 46
Lactuca
  biennis, 72
  canadensis, 72
    var. canadensis, 72
    var. latifolia, 72
    var. longifolia, 72
    var. obovata, 72
  floridana, 72

erectum, 32
hydropiper, 32
hydropiperoides, 32
    var. hydropiperoides, 32
    var. setaceum, 32
lapathifolium, 32
orientale, 32
pensylvanicum, 32
    var. eglandulosum, 32
    var. pensylvanicum, 32
perfoliatum, 32
persicaria, 32
punctatum, 32
    var. *robustius*, 32
ramosissimum, 32
robustius, 32
sachalinense, 32
sagittatum, 32
scandens, 32
    var. cristatum, 32
    var. dumetorum, 32
    var. scandens, 32
setaceum, 32
tenue, 32
virginianum, 32
Polymnia
  canadensis, 73
  uvedalia, 73
POLYPODIACEAE, 20
Polypodiales, 18
Polypodiophyta, 18
Polypodiopsida, 18
Polypodium
  appalachianum, 20
  polypodioides, 20
    var. *michauxianum*, 20
  virginianum, 20
  appalachianum × virginianum, 20
Polystichum
  acrostichoides, 20
Pontederia
  cordata, 90
PONTEDERIACEAE, 90
Populus
  alba, 36
  balsamifera, 36
    var. *subcordata*, 36
  × *barnesii*, 36
  × canadensis, 36
  × canescens, 36
  deltoides, 36
    subsp. *deltoides*, 36
    subsp. *monilifera*, 36
    var. deltoides, 36
    var. *missouriensis*, 36
    var. occidentalis, 36

× *gileadensis*, 36
grandidentata, 36
heterophylla, 36
× jackii, 36
nigra, 36
    var. italica, 36
× smithii, 36
tremuloides, 36
alba × tremula, 36
balsamifera × deltoides, 36
deltoides × nigra, 36
grandidentata × tremuloides, 36
Porteranthus
  stipulatus, 44
  trifoliatus, 44
Portulaca
  grandiflora, 30
  oleracea, 30
PORTULACACEAE, 30, 99
Potamogeton
  amplifolius, 76
  berchtoldii, 76
  crispus, 76
  diversifolius, 76
  epihydrus, 76
    var. *nuttallii*, 76
    var. *ramosus*, 76
  filiformis, 76
    var. alpinus, 76
    var. *borealis*, 76
  foliosus, 76
    var. *macellus*, 76
  friesii, 76
  gramineus, 76
    var. *maximus*, 76
    var. *myriophyllus*, 76
  × hagstroemii, 76
  hillii, 76
  illinoensis, 76
  natans, 76
  nodosus, 76
  pectinatus, 76
  perfoliatus, 76
    subsp. *bupleuroides*, 76
    var. bupleuroides, 76
  praelongus, 76
  pulcher, 76
  pusillus, 76
    var. *tenuissimus*, 76
  × rectifolius, 76
  richardsonii, 76
  robbinsii, 76
  spirillus, 76
  strictifolius, 76
  tennesseensis, 76
  vaseyi, 76

zosteriformis, 76
gramineus × richardsonii, 76
nodosus × richardsonii, 76
praelongus × richardsonii, 76
POTAMOGETONACEAE, 76
Potentilla
  anserina, 44
  argentea, 44
  arguta, 44
  canadensis, 44
  canescens, 44
  fruticosa, 44
  inclinata, 44
  intermedia, 44
  norvegica, 44
  palustris, 44
  paradoxa, 44
  pensylvanica, 44
  recta, 44
  reptans, 44
  simplex, 44
Prenanthes,
  alba, 73
  altissima, 73
  aspera, 73
  crepidinea, 73
  racemosa, 73
  serpentaria, 73
  trifoliata, 73
PRIMULACEAE, 40
Primulales, 40
Proboscidea
  louisianica, 64
Prosartes
  lanuginosa, 91
  maculata, 91
Proserpinaca
  palustris, 48
    var. crebra, 48
Proteales, 48
Prunella
  vulgaris, 60
    var. *lanceolata*, 60
Prunus
  americana, 44
    var. *lanata*, 44
  angustifolia, 100
  avium, 44
  cerasifera, 44
  cerasus, 44
  domestica, 44
    var. *insititia*, 44
  hortulana, 44
  mahaleb, 44
  mexicana, 44
  munsoniana, 44

var. *illinoense*, 66
perfoliatum, 66
Triphora
trianthophora, 93
var. *schaffneri*, 93
Triplasis
purpurea, 89
Tripsacum
dactyloides, 89
Trisetum
pensylvanicum, 89
Triticum
aestivum, 89
Trollius
laxus, 25
Tsuga
canadensis, 22
Tulipa
gesnerana, 91
Tunica
prolifera, 31
saxifraga, 31
Tussilago
farfara, 74
Typha
angustifolia, 90
× glauca, 90
latifolia, 90
angustifolia × latifolia, 90
TYPHACEAE, 90
Typhales, 89

ULMACEAE, 26, 100
Ulmus
alata, 100
americana, 26
minor, 26
procera, 26
pumila, 26
rubra, 26
serotina, 100
thomasii, 26
UMBELLIFERAE, 54, 97
Uniola
latifolia, 84
Urtica
chamaedryoides, 27
dioica, 27
subsp. *gracilis*, 27
var. dioica, 27
var. *gracilis*, 27
var. procera, 27
urens, 100
URTICACEAE, 27, 100
Urticales, 26
Utricularia

cornuta, 64
geminiscapa, 64
gibba, 64
intermedia, 64
macrorhiza, 64
minor, 64
ochroleuca, 99
vulgaris, 64
Uvularia
grandiflora, 91
perfoliata, 91
sessilifolia, 91

Vaccaria
hispanica, 31
Vaccinium
altomontanum, 40
angustifolium, 39
atrococcum, 39
brittonii, 39
caesium, 40
corymbosum, 39
lamarckii, 39
macrocarpon, 39
myrtilloides, 39
oxycoccos, 39
pallidum, 40
simulatum, 39
stamineum, 40
var. *melanocarpum*, 40
var. *neglectum*, 40
vacillans, 40
Valeriana
ciliata, 66
edulis, 66
subsp. *ciliata*, 66
var. ciliata, 66
officinalis, 66
pauciflora, 66
uliginosa, 66
VALERIANACEAE, 66
Valerianella
chenopodiifolia, 66
locusta, 66
radiata, 66
umbilicata, 66
Vallisneria
americana, 75
Veratrum
viride, 91
woodii, 90
Verbascum
blattaria, 63
forma blattaria, 63
forma erubescens, 63
phlomoides, 63

phoeniceum, 63
thapsus, 63
virgatum, 63
Verbena
bracteata, 59
canadensis, 59
× engelmannii, 59
hastata, 59
× *hybrida*, 100
× moechina, 59
simplex, 59
stricta, 59
urticifolia, 59
hastata × urticifolia, 59
simplex × stricta, 59
VERBENACEAE, 58, 100
Verbesina
alternifolia, 74
helianthoides, 74
occidentalis, 74
virginica, 74
Vernonia
fasciculata, 74
gigantea, 74
missurica, 74
noveboracensis, 74
gigantea × noveboracensis, 74
Veronica
agrestis, 63
americana, 63
anagallis-aquatica, 63
arvensis, 63
austriaca, 63
subsp. *teucrium*, 63
beccabunga, 63
catenata, 63
chamaedrys, 63
filiformis, 63
hederifolia, 63
latifolia, 63
longifolia, 63
officinalis, 63
peregrina, 63
subsp. *peregrina*, 63
subsp. *xalapensis*, 63
var. peregrina, 63
var. xalapensis, 63
persica, 63
polita, 63
scutellata, 63
serpyllifolia, 63
teucrium, 63
verna, 63
Veronicastrum
virginicum, 63
Viburnum

octoflora, 89
 var. glauca, 89

Waldsteinia
 fragarioides, 45
Wisteria
 floribunda, 48
 frutescens, 48
 macrostachya, 48
Wolffia
 borealis, 77
 brasiliensis, 77
 columbiana, 77
 papulifera, 77
 punctata, 77
Wolffiella
 floridana, 77
 gladiata, 77
Woodsia
 ilvensis, 20
 obtusa, 20
Woodwardia

areolata, 19
virginica, 19

Xanthium
 spinosum, 74
 strumarium, 74
  var. canadense, 74
  var. glabratum, 74
Xanthorhiza
 simplicissima, 25
XYRIDACEAE, 77, 101
Xyris
 caroliniana, 77, 101
 difformis, 77
 torta, 77

Yucca
 filamentosa, 91

Zannichellia
 palustris, 76
ZANNICHELLIACEAE, 76

Zanthoxylum
 americanum, 53
Zea
 mays, 89
Zelkova
 serrata, 26
Zigadenus
 elegans, 91
  subsp. *glaucus*, 91
  var. glaucus, 91
 glaucus, 91
Zinnia
 elegans, 74
 violacea, 74
Zizania
 aquatica, 89
Zizia
 aptera, 55
 aurea, 55
Zosterella
 dubia, 90
ZYGOPHYLLACEAE, 53

INDEX TO COMMON NAMES

CONTRIBUTORS

Barbara Kloha Andreas, a native of Dundee, Ohio, is Professor of Biology at Cuyahoga Community College and Adjunct Professor at Kent State University. Her areas of expertise are the ecology of Ohio's peatlands and the identification and distribution of Ohio's vascular plants and bryophytes. Andreas is a 1997 recipient of the Ohio Biological Survey's Herbert Osborn Award. She twice received "Paper of the Year Award" from The Ohio Journal of Science and has also received the President's National Stewardship Award from The Nature Conservancy. She is a member of the Ohio Flora Committee of The Ohio Academy of Science, the Ohio Rare Plants Advisory Committee of the ODNR's Division of Natural Areas and Preserves, and the Executive Committee of the Ohio Biological Survey. She is the author of numerous scientific papers and of a book, *The Vascular Flora of the Glaciated Allegheny Plateau Region of Ohio*. In addition, she is coauthor of two other books, *Floristic Index for Establishing Assessment Standards: A Case Study for Northern Ohio*, and *A Catalog and Atlas of the Mosses of Ohio*. Active in land preservation, she has served on the boards of The Wilderness Center and The Nature Conservancy and has been a Park Commissioner for Portage County, Ohio.

Tom S. Cooperrider is Emeritus Professor of Biological Sciences at Kent State University, where he taught for thirty-five years. A native of Licking County, Ohio, he graduated from Denison University before going on to graduate work at the University of Iowa. His M.S. and Ph.D. theses, both done under the direction of Robert F. Thorne, involved research on the flora of Iowa. Both were subsequently published by the University of Iowa. He and his wife, Miwako K. Cooperrider, also a student of Thorne's, built the Kent State Herbarium from 1,500 to over 60,000 specimens. Most of the specimens were collected from Ohio, many by some twenty graduate students. His *The* Dicotyledoneae *of Ohio, Part 2*, was published by OSU Press in 1995. He was the dedicatee in 1989 of Andreas's book, *The Vascular Flora of the Glaciated Allegheny Plateau Region of Ohio*, and in 1995 of the ODNR's Kent Bog State Nature Preserve. In 1994, he received the Herbert Osborn Award from the Ohio Biological Survey, of which he is a life member. He is a Fellow of The Ohio Academy of Science, The Explorers Club, and the American Association for the Advancement of Science.

Allison W. Cusick has been botanizing in Ohio for more than thirty years. In the eighth grade in Jefferson County, Ohio, he was introduced to biology by a remarkable teacher, Forest W. Buchanan. As an undergraduate at The Ohio State University, Cusick came under the influence of Clara G. Weishaupt of the Botany Department. After graduating from OSU with an M.A. in English, he began graduate work under Tom S. Cooperrider at Kent State University, where he earned an M.S. in botany. He has been employed by the Division of Natural Areas and Preserves, Ohio Department of Natural Resources, since 1978 and is presently Chief Botanist for the Division. He is the author or coauthor of more than forty scientific papers on the botany of Ohio and adjacent states. With Gene M. Silberhorn, he published a vascular flora of unglaciated Ohio which remains the standard reference on that area. He received the Herbert Osborn Award from the Ohio Biological Survey in 1998, in recognition of research, teaching, and service related to the Ohio flora.

Guy L. Denny is a respected interpretive naturalist, writer, and photographer with a wealth of knowledge about the natural history of Ohio. He retired in 1999 from his position as Chief of the Ohio Department of Natural Resources' Division of Natural Areas and Preserves after a career spanning thirty-three years of public service. A native of Toledo, Denny began his professional career as a naturalist with Toledo MetroParks, then became a teacher-naturalist for the Willoughby-Eastlake School System. He joined the State of Ohio workforce in 1969 as Chief Naturalist for the ODNR's Division of Parks and Recreation, later transferring to the ODNR's Division of Natural Areas and Preserves in 1976 as Assistant Chief of the Division. He is the author of several booklets on Ohio natural history, including *Ohio's Reptiles*, *Ohio's Trees*, and *Ohio's Amphibians*, and is past editor of the DNAP Newsletter, in which he published numerous natural history articles. Denny has served on several boards, including those of The Ohio Academy of Science, Ohio Alliance for the Environment, Ohio League of Conservation Voters, and the Outdoor Writers of Ohio. He currently serves as Secretary-Treasurer of the Outdoor Writers of Ohio and leads field trips and conducts slide presentations for numerous conservation organizations throughout the state.

John V. Freudenstein is Director of the Herbarium and a faculty member in the Department of Evolution, Ecology and Organismal Biology at The Ohio State University. A native of Michigan, he received a B.S. in botany at the University of Michigan, where he assisted Edward G. Voss in the preparation of the second volume of the *Michigan Flora*. He earned his Ph.D. in plant systematics from Cornell University. Freudenstein has held research positions at the University of Copenhagen, Harvard University, and the Royal Botanic Gardens, Kew, and is a former member of the faculty of biological sciences at Kent State University. His research interests center on phylogenetic systematics of the Orchidaceae, based on studies of morphological and molecular data.

John J. Furlow, who grew up in southern and central Indiana, has been a long-time resident of Ohio and an avid student of Ohio's natural history. He is widely known as a taxonomic specialist on the birch family. As a graduate student, he prepared a monograph of the American members of the genus *Alnus* in that family. He is currently studying the "dwarf" birches, a group of shrubby plants native to the circumpolar subarctic and cool temperate region of the Northern Hemisphere. He is preparing Part 1 of Volume 2 of *The Vascular Flora of Ohio*, which contains many of Ohio's most familiar tree species, and is revising E. Lucy Braun's well-known *Woody Plants of Ohio*. As curator of The Ohio State University Herbarium, a position he has held for nearly twenty years, Furlow oversees Ohio's largest collection of botanical specimens and performs a wide variety of public service activities. He is a popular speaker on topics ranging from the biology of trees, to Ohio's floristics and phytogeography, to various aspects of gardening and horticulture, and to botanical exploration in Latin America, where he has traveled and collected extensively.

John T. Kartesz was born and raised in McKeesport, Pennsylvania, a small suburb of Pittsburgh. After conducting undergraduate and graduate research at West Virginia University and The Ohio State University, and completing his Ph.D. at the University of Nevada, he assumed his current position as Director of the Biota of North America Program of the North Carolina Botanical Garden. Over the past thirty years he has gained international recognition for his North American floristic research. In 1980, he and his sister Rosemarie completed *A Synonymized Checklist of the Vascular Flora of the United States, Canada, and Greenland*. In 1994 he updated this work, which now serves as a national standard for vascular plant nomenclature and taxonomy. His most recent work, the *Synthesis of the North American Flora*, jointly authored with Christopher A. Meacham, represents the first digital floristic summary of North American vascular plants north of Mexico. Kartesz currently serves on the Flora of North America Management and Editorial Committees and is active in numerous other floristic and conservation endeavors.